The Causes of Evolution

Primula Floribunda and *Primula Verticillata*, the parents of *P. Kewensis*

The sterile hybrid *Primula Kewensis* and its fertile form produced by a doubling of the chromosome number. Is it a new species?

The Causes of Evolution

By J. B. S. Haldane, F.R.S.

With a new Introduction and Afterword
by Egbert G. Leigh, Jr.

PRINCETON UNIVERSITY PRESS
PRINCETON, NEW JERSEY

Published by Princeton University Press, 41 William Street, Princeton,
New Jersey 08540

Copyright © by J. B. S. Haldane. Preface and Afterword for the
Princeton Science Library edition © 1990 by Princeton University Press

Library of Congress Cataloging in Publication Data

Haldane, J.B.S. (John Burdon Sanderson), 1892–1964.
  The causes of evolution/by J.B.S. Haldane; with a new Preface and
Afterword by Egbert G. Leigh, Jr.
    p.   cm.—(Princeton science library)
  Reprint. Originally published: London; New York; Longmans,
Green, 1932.
  Includes bibliographical references.
  ISBN 0-691-02442-1
  1. Evolution. I. Title. II. Series.
QH366.H45   1990
575—dC20   89-77434

      ISBN 0-691-02442-1

    First Princeton Science Library printing, 1990

This book was first published by Longmans, Green & Co. Limited, in
1932; it was first published in paperback in the United States by Cornell
University Press in 1966; and is reprinted here, with new material, by
arrangement with Harper & Row, Inc. and Longman Group UK
Limited.

Editorial and production services by Fisher Duncan, 10 Barley Mow
Passage, London W4 4PH, UK

Second printing, with corrections, 1993

Printed in the United Kingdom

              10 9 8 7 6 5 4 3 2

Contents

Preface

This book is based on a series of lectures delivered in January 1931 at the Prifysgol Cymru, Aberystwyth, and entitled "A Re-examination of Darwinism." These lectures were endowed by the munificence of the Davies family, with the provision that their substance should be published in book form. This admirable condition ensures that, unlike the average university lectures, which stale with great rapidity, they should only be delivered once, and also that they should be made generally available before any novelty which they may possess has worn off.

Apart from the Appendix, I have added very little to the lectures as delivered. I doubt whether the time is yet ripe for a really comprehensive book covering the same ground, because our knowledge of the cytological nature of differences between species is increasing so rapidly as to render any account of these differences very provisional.

Readers who are not versed in biology will be well advised to skim lightly over Chapters II and III, which summarise our knowledge of those branches of genetics which are most important to the book's argument. I fear that the mathematical appendix will have a limited appeal. But I venture to hope that certain arguments in the body of the book (in particular that which purports to prove that mutation, Lamarckian transformation, and so on, cannot prevail against natural selection of even moderate intensity) will not be rejected unless a fallacy is discovered in the mathematical reasoning on which they rest.

I have to thank my colleagues of the John Innes Horticultural Institution, not only for permission to mention their unpublished work, but for many of the ideas which are here presented. Finally I wish to record here the very pleasant memories which I preserve of the week during which I had the honour to be a member of the staff of the National University of Wales.

J. B. S. Haldane

Preface

It has been a most extraordinary experience to be hailed forth from the baroque beauty of a tropical forest to introduce one of the three "founding classics" of the mathematical theory of evolution to a new generation of readers. I am ashamed to confess that this summons to reread Haldane brought home to me for the first time how clearly he foresaw modern issues of evolutionary theory. I am most grateful to the editors of Princeton University Press for their invitation.

The habit of publishing collections of papers of distinguished investigators—such a blessing to mathematicians and physicists—has yet to take hold in biology. Provine has performed this service for Sewall Wright, and Bennett for Ronald Fisher, to the great profit of modern biologists, but no one has yet done the same for Haldane. I have had to depend on the librarians of the Smithsonian Tropical Research Institute, and the editorial staff of Princeton University Press, for procuring copies of some of Haldane's papers for me. I am most grateful for their indulgence.

I am deeply indebted to Henry Horn and James Crow for encouragement of the most heartening kind, to Allen Herre and John Maynard Smith for reading the Introduction with a critical eye, to David Stern for reading the Afterword and suggesting useful improvements, and to Michael Turelli for casting an eye over section 8 of the Afterword and bringing a crucial reference to my attention. Finally, I give thanks for the trees and the animals of this tropical island, which help me put mathematical theories in proper and suitable perspective.

Egbert Giles Leigh, Jr.
Barro Colorado Island, Panama
Feast of All Saints, 1989

Introduction by Egbert G. Leigh, Jr.

FISHER, WRIGHT, HALDANE, AND THE RESURGENCE OF DARWINISM

Within a brief span in the early 1930s, Fisher (1930) and Haldane (1932) wrote books and Wright (1931) an extensive summary paper, all of which used mathematics to marry the new science of genetics to the older one of evolutionary theory. These three works gave a new lease of life to Darwin's theory of evolution. Before 1930, most biologists accepted evolution as a fact, but many of them doubted the capacity of natural selection of random mutations (random changes in the hereditary constitutions of organisms) to account for observed degrees of adaptation (see, for example, Berg 1926). Meanwhile, at least in the United States, "creationists" were fighting court battles to prevent the teaching of evolution in public schools. After Fisher, Wright, and Haldane refounded evolutionary theory on the principles of genetics, the anti-Darwinian tide was completely reversed. Soon, a steady procession of famous books—Dobzhansky (1937), Huxley (1942), Mayr (1942), and Simpson (1944) among them—used our growing understanding of genetics to illuminate the causes of evolution, thus creating the "neo-Darwinian synthesis of evolutionary theory."

To understand Haldane's contribution to this revolution in evolutionary theory, one must consider the immediately preceding works of Fisher (1930) and Wright (1931). These three works are wonderfully different. Fisher wished to establish whether natural selection of random mutations could and did account for observed degrees of adaptation. To this end he constructed a systematic, testable theory of evolution by natural selection, and followed where it led. His theory predicted that, especially in large populations, even slight selective differentials sufficed to substitute a favored gene for a less advantageous counterpart. He concluded accordingly that most organisms were very precisely adapted to their ways of life, and that large populations played a dominant role in the evolution of new adaptations. Since natural selection acts, in first instance, upon individuals, Fisher predicted that evolution hinges on what benefits individuals rather than populations or

References to Haldane (1932) in the Introduction give page numbers of this reprinted edition.

ix

species. Fisher was well aware of the conflicts that can arise between the good of human societies and the advantage of some of their individuals. He even suggested that the advantage to one's social advancement of having fewer children, accruing to abler individuals in civilized societies, results in a natural selection against ability and intelligence that has ruined whole civilizations. Nevertheless, Fisher paid no attention to how or why species are organized into relatively harmonious communities. Fisher treated genes simply as "black boxes" whose inner workings were of no concern, although he did assume that, by and large, "good genes make good genotypes," that the fitness of a gene combination bore some correlation with the sum of the average fitnesses of its component genes. In general, Fisher had no eyes for what conflicted with his theory, and no interest in what was irrelevant to it. Instead, he built a System. Despite its many faults and problems, Fisher's is the most useful and workable systematization of evolutionary theory yet devised. It is a model of what evolutionary theory should aim for, and a foundation for accomplishing that aim.

Wright asked what circumstances would best favor the origin and spread of novel adaptations. He had a better feel than Fisher for the complexities of how genes act in programming the development and behavior of organisms. Like Haldane (see p. 53), Wright may already have sensed that genes prescribe chemical reactions, which presumably modify the effects of other genes (Wright 1934). Where Fisher assumed that adaptation was best attained by testing each mutation one by one, on its average merits, Wright felt that the merits of a gene combination could not be predicted from the average merits of its component genes. Wright therefore argued that adaptive evolution is favored in species composed of many small, nearly isolated subpopulations. In such species, more different gene combinations would arise, primarily by chance, but a really advantageous gene combination could multiply and spread through the whole species. This theory could explain why so many species are composed of many, nearly isolated, subpopulations, and why differences between subpopulations, which presumably arose by chance, appear nonadaptive, while differences between families or orders of organisms are adaptive. Wright went on to construct his own system, rival to Fisher's. Wright's system was more appealing to many naturalists. It also provided rational objections to many of the assumptions crucial to Fisher's theory. However, Wright's theory is too complex, even now, to yield many useful predictions:

some of the difficulties involved are illustrated by Kauffman and Levin (1987).

Haldane did not build a system. He had a sharp eye for the unstated assumption, the hidden problem. Did his sense of the mysteries of life put him off system building? Remarks of his, which won grudging respect from one of the most partisan anti-Darwinians of our time (Jaki 1978, p. 294), suggest that this was indeed the case: "The world is full of mysteries. Life is one. The curious limitations of finite minds are another. It is not the business of evolutionary theory to explain these mysteries. Such a theory attempts to explain events of the remote past in terms of general laws known to be true in the present, assuming that the past was no more, but no less, mysterious than the present" (Haldane 1932, p. 3). Haldane's business was rather to weigh up Darwin's theory of evolution in the light of the new science of genetics. He was a superb storyteller, as his book for children and his essays attest (Haldane 1944, 1985). His *The Causes of Evolution*, originally a set of lectures, is much more readable than Fisher (1930) or Wright (1931) and attracted more attention, and more praise, when it first appeared. Haldane's book was, however, eclipsed later on by the systems of Fisher and Wright. These circumstances prompt three questions. What did Haldane's book contribute to the turning of the anti-Darwinian tide? What does its subsequent treatment tell us about the way science works? What can his book tell the modern student of evolution?

HALDANE AND THE RESTORATION OF DARWINISM

Haldane's book focuses on four general questions. First, are differences between species of the same nature as differences between individuals of the same species? Second, how does natural selection work, and how important a role does it play in evolution? Third, how can the obvious importance of natural selection be reconciled with the nonadaptive aspects of evolution? Finally, how does evolution reconcile the conflicting interests of gametes, individuals, and populations?

Can we infer "macroevolutionary" processes from "microevolution"?

To answer the first question, Haldane showed that the genetic differences between members of different species were of the same sort as the genetic differences between members of the same species (Haldane 1932, chap. 3). As Haldane remarked, the number of differences between members of different species are greater, but differences both between and within species can be explained largely in terms of differences between individual genes at comparable loci, differences between the arrangement of genes on chromosomes, and differences in number, either of individual chromosomes (aneuploidy) or of whole sets of chromosomes (polyploidy). Cytoplasmic differences do play a role, but a lesser one. Haldane took these facts to mean that the same processes that account for "microevolutionary" changes within populations also account for the differences involved in the evolution of new species. Drawing as he did on firmer and more abundant evidence, Dobzhansky (1937) was to hammer home this point much more forcefully. The methods of molecular genetics, which allow us to compare homologous genes in different organisms without performing crosses, have amply confirmed the conclusions of Haldane and Dobzhansky (see, for example, Maynard Smith 1989). The evidence from the genetic bases of the differences between species, showing that what we learn about microevolution is relevant to macroevolution, is one of the firmest buttresses of modern evolutionary theory.

How powerful is natural selection?

To answer this question, Haldane presented several examples of natural selection in the wild. Haldane discussed several instances where predators who eat the most conspicuously visible individuals of a species of insect have created an effective natural selection for cryptic coloration. Such studies of natural selection in the wild have been profitably extended by Ford (1971), Kettlewell (1973), and Endler (1986), among others. Haldane also sought to demonstrate, as Fisher (1930) had before him, that natural selection was by far the most efficacious agent for change in the genetic composition of a population. Haldane provided empirical evidence that mutation, unaided by selection, was a much weaker agent for change. He also cited Fisher's theoretical calculations showing that, at least in large populations, the effects of chance changes in the genetic composition of a population due to the accidents of who dies and who lives to reproduce are equally negligible compared to natural selection. Even while Haldane was writing this book, the famous dispute between Fisher and Wright erupted, concerning the role chance changes in the genetic composition of subpopulations plays in generating the appropriate "raw material" for natural selection. Fisher and Wright agreed, however, that natural selection is the ultimate "director" of evolution, and this is the prevalent, although not universal, opinion among evolutionary theorists today.

Haldane also commented on the sense in which natural selection can be said to "create." Not being God, natural selection cannot create *ex nihil*. Selection needs variation, deriving from mutations, on which to work. Nevertheless, mutations offer so many possible genotypes that natural selection, even though it is an automatic consequence of differential survival and reproduction, can be said to be a creative, organizing process (Haldane 1932, pp. 52–53), analogous in its effects to an author's arranging meaningful sequences of words from the multitudinous possibilities offered by the letters of the alphabet (Muller 1949).

Haldane, however, did not stop there. Like Fisher, he knew that most readily visible mutations are of unfavorable effect. How can one believe that mutations supply sufficient raw material for adaptation? To a large extent, Haldane accepted Fisher's argument that the larger a *random* change made in an organized system, the less likely that change is to be favorable, so that mutations of smaller effect are more likely to be favorable. If so, the deleterious nature of visible mutations does not imply that mutations of much smaller effect are so universally harmful. Haldane (1932, p. 89) argued, for example, that one could design two types of simple eye—one, like a camera, with a lens, and one, like a reflecting telescope, with a concave mirror. Haldane asserted that only the former occurs among animals, for only the former can evolve by small steps, each of which is advantageous. Organisms with eyes like reflecting telescopes have since been discovered: it is not quite as difficult for organisms to evolve mirrors as Haldane thought (Land 1978). Haldane recognized, however, that many species of plants have evolved by a single step, allopolyploidy, whereby, as a result of hybridization between two species, a new species is formed which possesses all the chromosomes of each parent species. Unlike Darwin and Fisher, Haldane refused to argue that evolution is necessarily a continuous process, invariably free of jumps: the evidence wouldn't let him do so. Whether macromutations—mutations of large, "systemwide" effect—play an important role in evolution is still much disputed (Løvtrup 1976; Leigh 1987). Nevertheless, recent analyses of the genetic differences between individuals of related species reinforce Haldane's conclusion that, aside from allopolyploidy, one species usually evolves from another by a succession of many small genetic changes rather than a few large ones (Lande 1981a). When evolution proceeds by small steps, one can speak most convincingly of the creative powers of natural selection.

Interestingly, Haldane refused to credit natural selection with creating mind, even though he insisted that mind evolved.

> . . . it is to be expected that the types of material system associated with mind, and hence the types of mind possible, will be severely restricted. We shall not be surprised to find considerable similarities between minds which have developed on quite different lines of descent. We shall not regard it as a mere coincidence that man cannot merely sympathize with the bee's devotion to its hive, but with its preferences regarding the colour and smell of flowers, and with its habit of dancing when it has satisfied its desires . . .
>
> If we are to have mind at all, it must probably conform to certain laws. There is no need to suppose that these laws, any more than those of biochemistry, are products of natural selection. Selection no doubt accounts for certain details, but in all probability not for the general character of mind. [Haldane 1932, p. 87]

In short, Haldane viewed the evolution of mind as directed by physical constraints. In fact, organisms have discovered optimal solutions, reflected in simple and elegant forms, to many of the physical problems posed by the various challenges of making a living (Vogel 1988). Are we to say that, in these respects, evolution has been directed by physical constraints? The sense, if any, in which physical or developmental constraints "direct" evolution has become the subject of lively discussion (Thomson 1985), in which the name of D'Arcy Thompson is often mentioned (Gould 1980a). Thompson (1942) thought that many biological structures arose, not as inherited adaptations, but as direct responses to physical forces: he viewed cell shapes, for example, as the product of surface tension. We must remember that the mechanical efficiency of form can be explained in two ways:

> If any given gene combination produces a structure of good mechanical design, it may, because of this . . . remain in the population through favourable selection. In other instances, there is clearly a direct effect of the environment which causes, by mechanical or physical forces, the forms of a living structure. In the case of these direct adaptations, we may assume that such responsiveness to the

environment is adaptively advantageous and, therefore, the gene
complement that favours responsiveness . . . to these environmental
conditions is . . . maintained by selection. [Bonner 1961, p. xi]

Selection is an agency that relates organisms to their environments,
physical and biological. Of course the environment affects the
direction of evolution. One must remember, nonetheless, that
engineers are viewed as creative if they use the laws of physics to
find new and better solutions to old problems. To deny the creativity of natural selection in connection with the evolution of mind
was to anticipate a semantic nightmare, but to concern oneself, as
Haldane did, with the mechanical and developmental context
within which natural selection operates is absolutely necessary.

One implication of Haldane's view of mind is parenthetically
worth noting, even if it takes us off our subject. If, as Haldane
thought, our mind works like those of other animals, then Polanyi
(1962) was quite right to insist that

> the meaning of an animal's actions can be understood only . . . by
> reading its mind in terms of these actions . . . and not by observing
> these actions themselves as we may observe inanimate processes.
>
> Behaviourists teach that in observing an animal we must refrain
> above all from trying to imagine what we would do if placed in the
> animal's position. I suggest, on the contrary, that nothing at all could
> be known about an animal that would be of the slightest interest . . .
> to psychology except by identifying ourselves with a center of action
> in the animal and criticizing its performance by standards set up for it
> by ourselves. [Polanyi 1962, p. 364]

It might surprise some that a materialistic view of mind could serve
as a basis for an intuitive, empathic approach to animal behavior.

Why does evolution so often seem nonadaptive?

Given the creativity and power of natural selection, Haldane
wondered why evolution so often seems nonadaptive or even
maladaptive, and he went to some trouble to explain how this could
be so. He remarked on the apparent nonadaptiveness of the
characteristics by which taxonomists distinguish related species
(1932, p. 62). He mentioned orthogenetic trends leading to extinction, such as the tendency of lineages of the oysterlike bivalve
Gryphaea to evolve more and more tightly coiled forms until, just
before extinction, the valves could barely open (1932, pp. 12–13).

And he noticed that many lineages evolved large size or large horns just prior to extinction (1932, p. 76). Haldane provided four classes of explanation for this puzzle. He viewed many unadaptive characteristics as correlative byproducts of invisible adaptations such as disease resistance (1932, p. 76) as Darwin did before him. He considered the possibility that certain structures were useful only during infrequent catastrophes that killed individuals lacking those structures (1932, p. 64). He was fascinated (1932, pp. 110–113) by some mathematics of Fisher which suggested that, under certain circumstances, a quantitative characteristic acted on by many genes could evolve past its selective optimum: a suggestion that turned out to be misleading, as we shall see in the Afterword. Finally, Haldane thought that, thanks to social or sexual competition among their members, common species were more liable to extinction than rare ones. He believed that, in a rare or scattered species, natural selection will make organisms fitter for their environment, for a rare species "is only engaged in competing with other species, and in defending itself against inorganic nature." On the other hand,

> as soon as a species becomes fairly dense matters are entirely different. Its members inevitably begin to compete with one another. . . . And the results may be biologically advantageous for the individual, but ultimately disastrous for the species. The geological record is full of cases where the development of enormous horns and spines (sometimes in the male sex only) has been the prelude to extinction. It seems probable that in some of these cases the species literally sank under the weight of its own armaments. [Haldane 1932, p. 65].

He commented on the "striking fact that none of the extinct species, which, from the abundance of their fossil remains, are well known to us, appear to have been in our own ancestral line. Our ancestors were mostly rather rare creatures. 'Blessed are the meek, for they shall inherit the earth' " (1932, p. 124). As this passage shows, Haldane thought the penalty of overabundance was indeed quite severe.

Today, the nonadaptive aspects of evolution seem less problematic. The days when differences between related species were considered nonadaptive have long since passed, sped on their way by studies of how these differences allow related species to coexist without one "outcompeting" the other (Pittendrigh 1950, 1958;

MacArthur 1958). What once seemed examples of suicidal orthogenesis or of undisciplined radiations of dying lines make more
sense, now that we have a better understanding of the biology of the
groups concerned. The conflict between individual advantage and
the good of the species, and the extent to which this conflict might
render evolution maladaptive, will be considered in the next section.

Conflicts between different levels of selection

Haldane knew that selection acts at different levels, and that there
may be conflicts between the advantage of gametes of individuals
and the good of the populations or species to which they belong. He
had no idea how these conflicts were resolved. Indeed, his interest
in these problems seems connected with explaining some apparently nonadaptive aspects of evolution. Plants compete to fertilize
the ovules of their species. Thus wind-pollinated species produce
vast amounts of pollen, far more than Haldane (1932, p. 66)
thought was necessary to pollinate the ovules of their species. He
thought that insect pollination, with its need for elaborate flowers
and rich nectar to attract the pollinators, was a maladaptive result of
the competition to pollinate ovules before one's neighbor does
(pp. 69–70). He also remarked that "there is competition between
pollen grains of the same plant on the basis of the genes carried by
them. . . . A gene which greatly accelerates pollen tube growth will
spread through a species even if it causes moderately disadvantageous changes in the adult plant" (pp. 66–67). As we have seen,
Haldane (pp. 65, 68–69), thought that the intense competition
between the members of a dominant or abundant species could
often favor characteristics that seriously impair the competitive
ability of the species and which might well cause its extinction. In
addition to showing how such "conflicts of interest" between
gamete and individual, or between individual and species, could
lead to evolution of maladaptive characteristics, Haldane (pp. 65–
66) used these observations to undermine the scientific basis of the
"social Darwinist" belief that unbridled competition between humans is good for humanity.

Although Fisher (1930) as well as Haldane emphasized potential
conflicts between an individual's advantage and the good of its
population, Fisher concluded that in such conflicts, individual
advantage nearly always prevails. The problem was subsequently
ignored until Wynne-Edwards (1962) proposed that selection be-

tween populations favors self-regulation in the numbers of each of these populations. Following Fisher's and Haldane's arguments, Williams (1966) demolished Wynne-Edwards's argument, but even he did not distinguish between the good of an individual and the advantage of one of its genes, or recognize the potential for conflict between the two. More recently, conflicts between various levels of selection have been a topic of lively discussion. Conflicts of interest between individuals and their genes (Crow 1979) or their cells (Buss 1987), an analysis of selection between species (Stanley 1975), and a recent summary of evidence that evolution somehow works for the good of the biosphere as a whole (Lovelock 1988) have all brought home the varieties of levels of selection and the opportunities for conflicts between them. Meanwhile, sexual selection, a topic long ignored, has become fashionable again (West Eberhard 1979). Transposable elements, genes which can jump from chromosome to chromosome and spread without regard to any benefits they might confer on their carriers, are now on center stage as potential exemplars of selfish genes (Engels 1986; Charlesworth 1988).

Why were biologists so slow to recognize the conflicts that Haldane foresaw, and that are now so topical? Are these conflicts rarer than Haldane thought? Can evolution reconcile conflicts between different levels of selection? To begin with, some of the examples that motivated Haldane to inquire into conflicts between individual advantage and the good of the species now appear inappropriate or ambiguous. Insect pollination is now thought to have been crucial to the adaptive radiation of flowering plants (Regal 1977). Wind pollination works for common plants, but the "directedness" of insect pollination is a necessity for rare ones. Populations of wind-pollinated conifers, for example, never consist entirely of scattered individuals. Even in otherwise diverse tropical forest, such conifers tend to be common where they occur (Regal 1977). Thus orchids require insect pollination because orchids are rare; not vice versa, as Haldane (1932, p. 69) had thought.

Despite Haldane's pessimistic view, even sexual selection may benefit the species. To begin with, sexual selection may make speciation easier. Two populations become separate species when the members of each population fail to recognize those of the other as suitable mates. Sexual selection can cause the mating preferences of members of different populations to diverge quite rapidly, thus transforming them into separate species (Lande 1981b; Eberhard 1985). As differential reproduction of species is an important aspect of selection between species, sexual selection may indeed be "good for the species."

Theory does speak with contradictory voices on whether sexual selection improves the "genetic quality" of a species. Kirkpatrick (1982) has modeled the effects of sexual selection on species where males contribute nothing but genes to their young. He showed that, if a male trait is sufficiently attractive to sufficiently many of the females, selection will strengthen female preference for that trait, even if the gene for that trait impairs the survival of both sexes to some degree. However, if a female's fertility depends on which mate(s) she chooses, selection favors "sensible" preferences, preferences which maximize the number of her offspring (Kirkpatrick 1985).

Empirical evidence, however, suggests that spectacular male traits, such as the peacock's tail or the male bowerbird's propensity for building enormous bowers, are not just "sensory traps" (West Eberhard 1984), structures which secure more matings by exploiting some innate responsiveness of females, regardless of their effect on other aspects of male fitness. Instead, these spectacular traits may often allow females to pass better-founded judgments on the prospective quality of their mates (Watt, Carter, and Donahue 1986; Diamond 1988; Poniankowski 1988). These traits may, however, render populations more susceptible to environmental change. Haldane may have been quite right to associate excessive response to sexual selection with susceptibility to extinction.

Can we conclude that, although conflicts between different levels of selection do exist, they are less obvious than Haldane's discussion might lead us to expect? Does evolution reconcile conflicts between different levels of selection?

The advantage of an individual and the good of its species should coincide more often than not, for a species where selection favors those individuals which render it less competitive or more susceptible to environmental change is a species which will probably disapper sooner than its competitors.

The means by which evolution reconciles such conflicts, however, have been more clearly elucidated at other levels. The problems involved recall those of framing a constitution which ensures that individual advantage coincides with the good of society. We have, however, the extra wrinkle that this "constitution" must itself be the product of natural selection. Consider, for example, a mutant gene which biases meiosis in its own favor, so that it passes to disproportionately many of its carriers' descendants. If this "segregation distorter" gene also disables its bearers somewhat, then a conflict of interest arises between this gene and its carriers. A gene on a different chromosome assorts independently of the distorter, and cannot spread itself by riding the distorter's coattails. On the other hand, the carriers of this second gene do suffer if they inherit the physical disability imposed by the distorter. A mutant of this gene accordingly can spread by suppressing the "segregation distortion," by restoring the honesty of meiosis—and thereby sparing some of its descendants' bearers the disability conferred by the distorter. In short, it is the common interest of an organism's genes to enforce the honesty of meiosis. In animal populations where individuals are clearly distinguishable, such as vertebrates or insects, the common interest of the genes has prevailed and meiosis is nearly invariably honest. No one has discovered how the few, exceptional distorter genes so far detected manage to cheat at meiosis, as if the manner of cheating is necessarily ingeniously devious. Conversely, if meiosis is "honest," a chromosomal gene can only spread by benefiting its carriers.

Human intelligence has yet to design a society where free competition among the members works for the good of the whole. The powerful, or the clever, can always bend the rules in their own favor, and some invariably do so. Indeed, the genius of a von Neumann could not imagine an economic "game" where the common interest of all the players enforced rules ensuring that competition between the players invariably benefited the whole (von Neumann and Morgenstern 1944). What human intelligence has yet to design, and what von Neumann was unable to imagine, for human societies, natural selection has very nearly accomplished for genes within organisms. How the genetic system evolved that allowed natural selection to accomplish this is still a mystery (Leigh 1987).

The question raised by Haldane concerning the possibility of conflicts between different levels of selection has thus led to one of the most curious and interesting chapters in evolutionary theory. Wht other surprises does the study of the interactions between different levels of selection hold in store for us?

Concluding remarks

The two things which impress me most about Haldane's book are its insight and its light-hearted refusal to be confined by a system. His insight has, I hope, been made clear. His third chapter outlines the road by which Dobzhansky and many others showed that differences *between* species are not of types qualitatively distinct from differences *within* species. This observation is the basis for believing that the same factors account for both microevolutionary differentiation within populations, and speciation. In his fifth chapter Haldane raises questions concerning conflicts between different levels of selection, answers to which have yielded the most convincing evidence yet of the importance of natural selection in evolution (reviewed in Leigh 1986). As a whole, Haldane's book balances theory with empirical observation in a most effective harmony. Haldane may have built no system, but he knew how to establish the foundations for understanding the role of natural selection in evolution. Today, biologists are again attacking Darwin's conception of natural selection as the "organizing influence" in evolution (Gould 1980b), and creationists once again deny the fact of evolution. Haldane's book is as appropriate a response to these doubts now as it was in 1932.

Haldane's freedom of mind is most evident in his last chapter. There he argues for a materialistic theory of mind, a "monism" as opposed to the traditional dualism of matter versus mind or spirit. Notice, however, the attitude with which he does so. He observes that "monism has the advantage that if it is wrong, it will ultimately lead to self-contradiction, whereas dualistic systems, which purport to give a less complete account of the world, are therefore less susceptible of disproof" (Haldane 1932, p. 84). This recalls the words of D'Arcy Thompson (1942, pp. 14–15) concerning the role of physics in biology: ". . . the appropriate and legitimate postulate of the physicist, in approaching the physical problems of the living body, is that with these physical phenomena, no alien influence interferes. But the postulate, though it is certainly legitimate, and though it is the proper and necessary prelude to scientific enquiry, may some day be proven to be untrue; and its disproof will not be to the physicist's confusion, but will come as his reward." Both passages reflect open-hearted inquiry, not commitment to a necessary truth. This openness sets Haldane apart from most of his fellow materialists, and from most of their religious opponents.

Haldane's freedom of mind permitted him some extraordinary insights. To be sure, he was no friend of Christian dogma. To begin with, the whole point of evolutionary theory is to provide a mechanistic (and therefore materialistic) account of the adaptation and diversification of living things, without appeal to miracle or divine intervention. Even though St. Thomas Aquinas insisted on the efficacy of "second causes", the freedom of objects and creatures to follow their own natures (Gilson 1983, pp. 180–184)—after all, "to deprive things of actions of their own is to belittle God's goodness," as well as to make natural science impossible—many viewed the extension of "second causes" to the origin of species as an infringement of divine prerogative. Perhaps in response to this tension, Haldane (1932, p. 86) wondered whether asserting that God made the tapeworm "fits in either with what we know of the process of evolution or what many of us believe about the moral perfection of God." Haldane also thought evolution did not much resemble a process governed by an intelligent Designer. He concluded accordingly (p. 86) that "at present it does not seem necessary to postulate divine or diabolical intervention in the course of the evolutionary process." In his very next sentence, however, he remarks that "The question whether we can draw theological conclusions from the fact that the universe is such that evolution has occurred in it is quite different, and very interesting." This question—one of Haldane's finest one-liners—has recently found its way to center stage (Barrow and Tipler 1986). A few pages later, moreover, he states that "My suspicion of some unknown type of being associated with evolution is my tribute to its beauty, and to that inexhaustible queerness which is the main characteristic of the universe which has impressed itself upon me during twenty-five years of scientific work." (pp. 90–91) The beauty, and the obviously contingent order, of the universe—hardly the product either of chance, or of laws discernible a priori which could not possibly have yielded any other result—are the foundation of the argument from "natural theology" for an idiosyncratic God (Mascall 1956). Haldane's is as good a précis of this argument as one could wish, not least in showing its limits, for this argument does not tell us very much about the God it infers.

WHAT DOES HALDANE'S BOOK TELL US ABOUT THE CONDUCT OF SCIENCE?

Haldane's book was eclipsed somewhat by the work of Fisher and Wright, apparently because Haldane refused to found a system.

Systems do have their advantages. Fisher's system gave him a prophetic faith in the adaptedness of organisms. This faith went far beyond the evidence of his day, but it has been triumphantly vindicated since. Similarly, Darwin's system convinced him that the fossil record would some day testify decisively for the fact of evolution, even though it had yet to do so. Nowadays, however, whatever "jumps" may remain in it, the fossil record allows us to speak unhesitatingly of evolution rather than a series of special creations.

Evolutionary theory, moreover, poses special problems that seem soluble only with the help of a system. Molecular models, drawing only on the known laws of mechanics, tell us that a molecular theory *can* account for the properties of gases. A self-contained mechanical model of continental drift may soon be available to demonstrate that the physical forces known to act in the earth's mantle and crust suffice to account for the rough periodicity (about 300 million years: Fischer 1984) with which continents aggregate into a Pangaea and break up again. As yet, however, no such model demonstrates that natural selection from random mutations can account for observed levels of adaptation, complexity and diversity (Leigh 1986). The radical inhomogeneity of living matter, at all levels of biological organization, may never permit it (Elsasser 1966). To test the notion of evolution by natural selection from random mutations, it appears we must frame a systematic theory of such aspects of the process as we can manage, and test such predictions as it yields; this was Darwin's path (Ghiselin 1969), and Fisher's, and it yields some measure of conviction to many minds, but in the absence of the appropriate "mechanical model" it is hard to convince a creationist.

More generally, we need a system of sorts to make any sense at all of the real world. Jolly (1985) showed the stages by which a child develops a system that enables it to make sense of its perceptions, that enables it to set each perception, as it were, in an appropriate context.

Nevertheless, systems can blind us both to what has no place in them and to what conflicts with them. Haldane raised questions concerning, among other things, conflicts between different levels of selection and the role of developmental constraints in evolution. These questions were passed by in the urgent rush to see if evolution by natural selection from random mutations can account for the basic phenomena of adaptation. Now that answers to them are needed for further progress in that direction, we find that Haldane's questions are indeed the appropriate next stage in learning how

adaptation can evolve. We are now ready to reap the benefit of the
fact that Haldane was a free man in the sense that really matters.
This book is witness to the riches of that freedom of thought.

REFERENCES

Barrow, J. D., and F. J. Tipler. 1986. *The Anthropic Cosmological
Principle*. Oxford: Oxford University Press.
Berg, L. S. 1926. *Nomogenesis, or Evolution Determined by Law*. London:
Constable.
Bonner, J. T. 1961. The editor's introduction. In D. W. Thompson, *On
Growth and Form* (abridged edition), pp. vii–xiv. Cambridge Eng.:
Cambridge University Press.
Buss, L. W. 1987. *The Evolution of Individuality*. Princeton, N.J.: Prince-
ton University Press.
Charlesworth, B. 1988. The maintenance of transposable elements in
natural populations. In *Plant Transposable Elements*, ed. O. Nelson, pp.
189–212. New York: Plenum Press.
Crow, J. F. 1979. Genes that violate Mendel's rules. *Scientific American*
240(2):134–146.
Diamond, J. 1988. Experimental study of bower decoration by the bower-
bird *Amblyornis inornatus* using colored poker chips. *American Natural-
ist* 131:631–653.
Dobzhansky, Th. 1937. *Genetics and the Origin of Species*. New York:
Columbia University Press.
Eberhard, W. G. 1985. *Sexual Selection and Animal Genitalia*. Cambridge,
Mass.: Harvard University Press.
Elsasser, W. M. 1966. *Atom and Organism*. Princeton, N.J.: Princeton
University Press.
Endler, J. A. 1986. *Natural Selection in the Wild*. Princeton, N.J.: Prince-
ton University Press.
Engels, W. R. 1986. On the evolution and population genetics of hybrid-
dysgenesis-causing transposable elements in *Drosophila*. *Philosophical
Transactions of the Royal Society of London* B312:205–215.
Fischer, A. G. 1984. The two Phanerozoic supercycles. In *Catastrophes and
Earth History*, eds. W. A. Berggren and J. A. van Couvering, pp. 129–
150. Princeton, N.J.: Princeton University Press.
Fisher, R. A. 1930. *The Genetical Theory of Natural Selection*. Oxford:
Oxford University Press.
Ford, E. B. 1971. *Ecological Genetics*. London: Chapman and Hall.
Ghiselin, M. T. 1969. *The Triumph of the Darwinian Method*. Berkeley:
University of California Press.
Gilson, E. 1983. *The Christian Philosophy of St. Thomas Aquinas*. New
York: Octagon Books.
Gould, S. J. 1980a. Double trouble. In *The Panda's Thumb*, ed. S. J.
Gould, pp. 35–44. New York: W. W. Norton & Co.
———. 1980b. Is a new and general theory of evolution emerging? *Paleobio-
logy* 6:119–130.

Haldane, J. B. S. 1932. *The Causes of Evolution*. London: Longmans Green & Co.

———. 1944. *My Friend, Mr. Leakey*. Harmondsworth, U.K.: Puffin Books.

———. 1985. *On Being the Right Size, and Other Essays*. Oxford: Oxford University Press.

Huxley, J. 1942. *Evolution, the Modern Synthesis*. London: George Allen and Unwin.

Jaki, S. L. 1978. *The Road of Science and the Ways to God*. Chicago: University of Chicago Press.

Jolly, A. 1985. *The Evolution of Primate Behavior*. New York: Macmillan.

Kauffman, S., and S. Levin. 1987. Towards a general theory of adaptive walks on rugged landscapes. *Journal of Theoretical Biology* 128:11–45.

Kettlewell, H. B. D. 1973. *The Evolution of Melanism*. Oxford: Oxford University Press.

Kirkpatrick, M. 1982. Sexual selection and the evolution of female choice. *Evolution* 36:1–12.

———. 1985. Evolution of female choice and male parental investment in polygynous species: The demise of the "sexy son." *American Naturalist* 125:788–810.

Land, M. F. 1978. Animal eyes with mirror optics. *Scientific American* 239(6):126–134.

Lande, R. 1981a. The minimum number of genes contributing to quantitative variation within and between populations. *Genetics* 99:541–553.

———. 1981b. Models of speciation by sexual selection on polygenic traits. *Proceedings of the National Academy of Sciences, USA* 78: 3721–3725.

Leigh, E. G., Jr. 1986. Ronald Fisher and the development of evolutionary theory I. The role of selection. *Oxford Surveys in Evolutionary Biology* 3:187–223.

———. 1987. Ronald Fisher and the development of evolutionary theory II. Influences of new variation on evolutionary process. *Oxford Surveys in Evolutionary Biology* 4:212–263.

Lovelock, J. 1988. *The Ages of Gaia*. New York: W. W. Norton & Co.

Løvtrup, S. 1976. On the falsifiability of neo-Darwinism. *Evolutionary Theory* 1:267–283.

MacArthur, R. H. 1958. Population ecology of some warblers of northeastern coniferous forests. *Ecology* 39:599–619.

Mascall, E. L. 1956. *Christian Theology and Natural Science*. London: Longmans Green & Co.

Maynard Smith, J. 198. *Evolutionary Genetics*. Oxford: Oxford University Press.

Mayr, E. 1942. *Systematics and the Origin of Species*. New York: Columbia University Press.

Muller, H. J. 1949. The Darwinian and modern concepts of natural selection. *Proceedings of the American Philosophical Society* 93:459–470.

Pittendrigh, C. S. 1950. The ecoclimatic divergence of *Anopheles bellator* and *Anopheles homunculus*. *Evolution* 4:43–63.

———. 1958. Adaptation, natural selection and behavior. In *Behavior and*

Evolution, eds. A. Roe and G. G. Simpson, pp. 390–416. New Haven: Yale University Press.

Polanyi, M. 1962. *Personal Knowledge*. Chicago: University of Chicago Press.

Pomiankowski, A. N. 1988. The evolution of female mate preferences for male genetic quality. *Oxford Surveys in Evolutionary Biology* 5:136–184.

Regal, P. J. 1977. Ecology and evolution of flowering plant dominance. *Science* 196:622–629.

Simpson, G. G. 1944. *Tempo and Mode in Evolution*. New York: Columbia University Press.

Stanley, S. M. 1975. A theory of evolution above the species level. *Proceedings of the National Academy of Sciences, USA* 72:646–650.

Thompson, D. W. 1942. *On Growth and Form*. Cambridge, Eng.: Cambridge University Press.

Thomson, K. S. 1985. Essay review: The relationship between development and evolution. *Oxford Surveys in Evolutionary Biology* 2:220–233.

Vogel, S. 1988. *Life's Devices*. Princeton, N.J.: Princeton University Press.

Von Neumann, J., and O. Morgenstern. 1944. *Theory of Games and Economic Behavior*. Princeton, N.J.: Princeton University Press.

Watt, W. B., P. A. Carter, and K. Donahue. 1986. Females' choice of "good genotypes" as mates is promoted by an insect mating system. *Science* 223:1187–1190.

West Eberhard, M. J. 1979. Sexual selection, social competition, and evolution. *Proceedings of the American Philosophical Society* 123:222–234.

———. 1983. Sexual selection, social competition, and speciation. *Quarterly Review of Biology* 58:155–183.

———. 1984. Sexual selection, competitive communication, and species-specific signals in insects. In *Insect Communication*, ed. T. Lewis, pp. 283–324. New York: Academic Press.

Williams, G. C. 1966. *Adaptation and Natural Selection*. Princeton, N.J.: Princeton University Press.

Wright, S. 1931. Evolution in Mendelian populations. *Genetics* 16: 97–159.

Wright, S. 1934. Physiological and evolutionary theories of dominance. *American Naturalist* 68:25–53.

Wynne-Edwards, V. C. 1962. *Animal Dispersion in Relation to Social Behavior*. Edinburgh: Oliver and Boyd.

CHAPTER I

Introduction

"Darwinism is dead."—*Any sermon.*

Seventy-two years have now elapsed since Darwin and Wallace (1858) formulated the theory that evolution had occurred largely as a result of natural selection. The doctrine of evolution was not, of course, new. But Lamarck and other eminent biologists had failed to convince the scientific world or the general public that evolution had occurred, still less that it had occurred owing to the operation of any particular set of causes. Darwin contrived to carry a considerable measure of conviction on both these points. The result has been that a generation ago most people who believed in evolution held that it had been largely due to natural selection. Nowadays a certain number of believers in evolution do not regard natural selection as a cause of it, but I think, that in general the two beliefs still go together.

So close a correlation is rather rare in the history of human thought. For example, men had been aware for ages of the existence of a past history of the human race before Daniel (or the author of the Book of Daniel) made the first attempt to view that history as a whole, and give a summary account of it. If Daniel had been the first person to persuade thinking men that the past had differed appreciably from the present, it is clear that his particular account of the historical process would have had a greater intellectual influence than it has actually had. We must therefore carefully distinguish between two quite different doctrines which Darwin popularised, the doctrine of evolution, and that of natural selection. It is quite possible to hold the first and not the second. Similarly with regard to the doctrines of Darwin's great contemporary Marx, it is possible to adopt socialism but not historical materialism.

Darwinism has been a subject of embittered controversy ever since its inception. The period up till Darwin's death saw a vast mass of criticism. This was mostly an attack on the doctrine of evolution, and was almost entirely devoid of scientific value. The few really pertinent attacks were lost amid a jabber of ecclesiastical bombinations. The criticism was largely dictated by disgust or fear of this doctrine, and it was natural that the majority of scientific men rallied to Darwin's support. By the time of Darwin's death in 1882, Darwinism had become orthodox in biological circles. The next generation saw the beginnings of a more critical attitude among biologists. It was possible to criticise Darwin without being supposed to be supporting the literal authenticity of the Book of Genesis. The criticism came from all sides. Palaeontologists, geneticists, embryologists, psychologists, and others, found flaws of a more or less serious character in Darwin's statements. But they almost universally accepted evolution as a fact.

The rising generation of biologists, to which I belong, may now perhaps claim to make its voice heard. We have this advantage at least over our predecessors, that we get no thrill from attacking either theological or biological orthodoxy; for eminent theologians have accepted evolution and eminent biologists denied natural selection.

In this course of lectures I do not propose to argue the case for evolution, which I regard as being quite as well proven as most other historical facts, but to discuss its possible causes, which are certainly debatable. It will, however, be worth while briefly to explain what is meant by evolution, and to indicate the arguments which lead the overwhelming majority of biologists to believe in it.

By evolution we mean the descent from living beings in the past of other widely different living beings. How wide the difference must be before the process deserves the name of evolution is a doubtful question. Many would refuse to dignify the changes which man has effected in the dog as evolution, though they have certainly an obvious bearing on the question of evolution. In the first generation after Darwin it was pointed out that artificially produced races, if they were incapable of breeding together, were so on mechanical grounds only, and never gave sterile hybrids like the mule. Since then races which, like species, are sterile on physiological grounds, or which give sterile hybrids, have been artificially produced; but those who have produced them are chary of claiming that they have originated a new species.

Certain of the critics of evolution have admitted the possibility of fairly large structural or functional changes, but not of such a profound change as the origin of consciousness or reason. I sympathise with their attitude, but cannot share it, because it seems to me to rest on a refusal to face certain perfectly amazing facts of everyday life. The strangest thing about the origin of consciousness from unconsciousness is not that it has happened once in the remote past, but that it happens in the life of every one of us. An early human embryo without nervous system or sense organs, and no occupation but growth, has no more claim to consciousness than a plant—far less than a jelly-fish. A new-born baby may be conscious, but has less title to rationality than a dog or ape. The evolutionist makes the very modest claim that an increase in rationality such as every normal child shows in its lifetime has occurred in the ancestors of the human race in the last few million years. He does not claim to be able to explain this process adequately, or even to understand it. But he claims that such an increase in rationality is a fact of everyday experience. It is conceivable, though to my mind unlikely, that there was a sharp break at some point in human evolution at which a new type of mental activity suddenly became possible. But there is a vastly greater probability of finding evidence of such a discontinuity in individual than in racial history. I do not think the likelihood very great, but if I believed in such radical changes, that is where I should be inclined to look for them.

This is, I think, a fair sample of the reply to a great many criticisms of any theory of evolution. The world is full of mysteries. Life is one. The curious limitations of finite minds are another. It is not the business of an evolutionary theory to explain these mysteries. Such a theory attempts to explain events of the remote past in terms of general laws known to be true in the present, assuming that the past was no more, but no less, mysterious than the present.

"Bist du gehemmt, das neues Wort dich stört,
Willst du nur hören, was du schon gehört,
Dich störe nichts, wie es auch weiter klinge,
Schon längst bekannt der wunderbarste Dinge."

While I shall not attempt to defend the historical side of the evolutionary theory, I propose to review the type of evidence on which it is based. First and foremost comes the evidence of fossils. Where a hundred years ago we had only small samples of a few populations at certain dates in the past, we have now in a few cases continuous records over enormous periods, and where the record is not continuous, very numerous different stages. Thus, thanks mainly to the work of Osborn and his colleagues, we now know of over 260 fossil species lying on or near the line of descent of the modern horse and its living relatives from four-toed and short-toothed ancestors. When one has made acquaintance with such series of related types any hypothesis other than evolution becomes fantastic.

If we had no fossil record at all, evolution would still be a plausible hypothesis to account for the structural relationships between living plants and animals, but there would often be a controversy as to whether certain simple forms were primitive or degenerate. Such disputes occur with reference to various groups of worms whose ancestors have left no fossils. A generation ago it was rather fashionable in such cases to support the hypothesis that certain simple forms were degenerate, just as Darwin's contemporaries had plumped for the opposing view. To-day we find that the older generation was often right. Thus the lamprey and its relatives the cyclostomes, fish-like vertebrates which have no lower jaw, were naturally regarded as representing a phase of vertebrate evolution earlier than the ordinary fish. This was confirmed by the brilliant work of Stensiö, who has recently shown that the Ostracodermi, a group of fish which was dominant about 400 million years ago when fish first appear in the geological record, possessed many features of internal anatomy characteristic of the lamprey.

Again comparative embryology has been of great value in tracing relationships. On the one hand the fact that every one of us, before birth, had at different times gill-slits, a tail, and a coat of hair, is barely intelligible on any hypothesis other than that of evolution. Moreover, apparently unrelated animal groups, such as molluscs and segmented worms, start life as embryos of the same type, and are therefore generally believed to have had a common ancestor before the fossil record began.

Geographical distribution again becomes intelligible if different plant and animal groups originated in different centres. Darwin and Wallace used such data to great effect, and more recently Willis (1922) and Vaviloff (1922) have drawn conclusions of great importance from the study of distribution. Roughly speaking, the older a group of organisms, the wider its distribution, apart from relict species on the verge of extinction. Relatively recent groups are usually restricted, e.g. the guinea-pig family to South America. In the same way, recently separated islands, such as England, have far fewer peculiar species than islands of long standing, such as New Zealand.

Of late years several new branches of comparative biology have been of value in working out relationships. For example, the majority of mammals are capable of oxidising uric acid to a more soluble substance, allantoin. Man is not; hence his is liable to gout. Most of the monkeys can oxidise uric acid. Our inability is shared by the tailless apes such as the gorilla and the chimpanzee. This fact certainly adds to the improbability of the view held in some quarters that man and the tailless apes sprang separately from tailed stock. Again, a study of blood transfusion shows that human bloods fall into four groups. A pint of my blood could be injected into any other man or woman with fair safety. A pint of most human bloods would kill me. I happen to belong to the group of so-called universal donors. There are three other groups, each with specific properties. The human blood groups are found in apes, such as the chimpanzee. Hence comes the paradoxical fact that it may be no more dangerous to have a transfusion of blood from a chimpanzee than from your own brother, let alone a lower animal. Innumerable facts of this kind go to show that the relationships between plants and animals indicated by the evolutionary hypothesis extend to chemical composition as well as structure.

Lastly, comparative parasitology supports the evolutionary hypothesis. If two animals have a common ancestor, their parasites are likely to be descended from those of the ancestor. This principle has been applied with considerable effect to the classification of frogs and other groups. Of course it does not imply that parisites cannot pass from one species to another. Thus that common enemy of man, the bed-bug, belongs to a family whose members are mostly parasitic on bats. Dr. Buxton has, I think, suggested that it is a relic of the association of our palaeolithic ancestors with bats in caves.

We must now consider some of the hypotheses which have been put forward to explain evolution. The fact to be explained is why one generation differs a little from its parents; why the average weight is slightly greater, the proportion of blue-eyed less, the average milk-yield greater, to take three possibilities. It is at once clear that some of these differences may be directly due to changed environment. Thus good feeding of cattle has a huge influence on their weight and milk yield. In order to allow for such effects it is desirable to compare two generations brought up in environments as similar as possible. In any case it is quite certain that changes of environment only produce notable effects on a species within a single generation in a very few cases. You can often make the progeny of a thin cow fat, or conversely, by good or bad feeding.

But there is no question that both are cows. In a few highly plastic species such as *Polygonum amphibium* or *Amblystoma tigrinum*, one can convert a land form into a water form, or conversely, and they differ structurally as much as do different species or even genera. In a few evolutionary series it is possible that continuous changes in environment may have played an important part. If in one or two generations we could have brought a group of Ammonities[1] from the sea water of Devonian times into a medium containing the salts of Cretaceous sea water their shells would probably have been altered slightly, and some of the evolutionary changes in Ammonites may have been due to causes of this kind. But it is clear that changes producible in the course of a few generations have been of a quite subordinate importance in evolution. We may now proceed to classify the causes which have been suggested for the deeper transformations shown by the geological record.

(*a*) Inheritable variations of an essentially random character. A good example is furnished by the colours of kittens in a mixed litter. We now know that variation of this kind is mainly due to the process of segregation, which will be described later. Taken by itself it will not explain evolution.

(*b*) Inheritable variation due to the action of the environment on the organism. It was once thought that all differences due to variations of environment were inherited. This was Lamarck's theory. We now know that this is not true; nevertheless since the time-scale of evolution is much longer than Lamarck supposed, a very slight tendency of characters so acquired to be inherited might have an important evolutionary effect.

(*c*) Variation due to internal causes, but not at random. It is thought by certain biologists that the lines of its future evolution are laid down in any organism, and that it will evolve on these predestined lines in spite of a variety of obstacles. An exposition of this point of view is to be found in Berg's "Nomogenesis," in my judgment by far the best anti-Darwinian book of this century. An English translation is available, though the original was published in Russian, in spite of the definitely Darwinian bias of the ruling group there. I wish to take this opportunity to acknowledge my indebtedness to Berg for several facts quoted in this book.

---

[1] Here and throughout I use the word "Ammonite" to denote a member of the Ammonoidea, and not in its more restricted sense.

In so far as they are formulated at all, most of the various theories which ascribe evolution to the guidance of an intelligent spirit or spirits should be grouped here. The Bishop of Birmingham, however, has recently suggested that variation is at random, God controlling evolution by means of the environment, and not from within.

(d) Variation due to hybridisation. This process may merely lead to new combinations of the characters of the species or groups which hybridise, but we shall see that it may also produce something entirely fresh.

(e) Selection. Darwin distinguished between natural and sexual selection. But the distinction is not fundamental. Thus we now know that the asexual workers of termites are attracted by the smell of the queen and apparently feed her for this reason, while wasp grubs repay the workers for their food by a drop of sweet secretion. These sensual attractions are clearly comparable to those which draw one sex to the other, and we shall see later that from the evolutionary point of view sexual selection is only one of a group of similar types of selection.

Darwin believed that selection acted on variations of types (a) and (b), i.e. random variations and inherited effects of use and disuse. Lamarck had attempted to ascribe all evolution to variations of type (b). Darwin saw that type (a) was more common, but also attached importance to type (b), especially to the inherited effects of use and disuse. R. A. Fisher, in his brilliant book, "The Genetical Theory of Natural Selection," to which I am much indebted, points out the reason for this fact. Darwin believed that the crossing of two types generally led to a blend, and that consequently bisexual reproduction tended to make a species uniform. He therefore had to postulate some cause constantly at work to keep up the inheritable variation within a species. He very naturally looked to the effects of differences of environment. It is clear that he was not comfortable about the matter. Thus he wrote to Huxley on November 25, 1859, "If, as I must think, external conditions produce little effect, what the devil determines each particular variation?"

Now Darwin's evidence as to blending came from crosses between lines of domesticated animals and plants which had been kept separate for considerable periods. If we cross members of a large and small race of poultry, the offspring are fairly uniform and intermediate in size. But this is not so if we continue the mating for several generations. The second generation of such a cross gives a great variety of sizes. If the blending has been permanent, as when water and ink are mixed, the second generation would be uniform. Actually in a population mating either at random or according to any law which is the same in every generation the amount of heritable variation is practically constant, apart from the effects of selection. The tendency to blending is exactly balanced by the opposite process of segregation, by which the offspring of a given union vary among themselves in respect of heritable characters. But whereas the phenomenon of heredity had been known in a general way for ages, that of segregation was first seriously studied by Mendel, in the nineteenth century, and it was above all Bateson who stressed its significance as a biological fact as important as heredity.

The amount of variation can in general only be altered by selection on the one hand and changes in the system of mating on the other. Thus if in man only persons over six feet high were allowed to have children the population would become taller on the average, and also more uniform. If incest were allowed and practised it would become more diverse owing to the appearance of monstrosities of many different kinds. Darwin observed blending, *i.e.* a diminution of variation, because the mating system of his domestic animals and plants was suddenly changed, for example, when two races of pigeons, which had been bred separately for many years, were mated together.

Further, though differences of environment do cause variation, this variation is not usually inherited to any measurable extent. This fundamental fact, which has been guessed at by Kant and others, was first demonstrated during the nineteenth century by the de Vilmorin family in France, and was part of the basis of the methods of selective breeding which they invented. But Louis de Vilmorin's "Notices sur l'amélioration des plantes par les semis et considérations sur l'hérédité dans les végétaux" was regarded as a mere practical handbook, and only runs to sixty pages (for a summary of his views see de Vilmorin, 1856). If, like most writers on heredity, he had gone beyond his facts, he would doubtless have attracted more attention. Weismann's statement of the principle was based on inadequate evidence, but his *a priori* arguments in its favour carried conviction in many quarters. His great service to science was his account of the behaviour of the chromosomes in connection with reproduction. The first really conclusive proof was given by Johannsen in 1903. He showed that when plants are self-fertilised for many generations the progeny of one of them forms what is called a pure line, in which differences are not inherited.

For example, Table I shows the non-inheritance of weight within a pure line of beans. In each generation Johanssen bred from light and heavy beans, and on the whole there was no resemblance between parents and offsprings as regards weight, as appears from the fact that the negative entries in the last column outweigh the positive. The weight is very susceptible to environment, as appears on comparing the weights in different years. But the changes produced by the environment are not inherited. Of course an ordinary population of beans consists of a number of pure lines, so selection is quite effective at first. But its ultimate result it to isolate the pure line with the largest or smallest mean weight. The same holds true for other characters and other organisms. Selection is effective during the first few generations, but sooner or later a pure line is generally reached, and selection becomes ineffective because the differences selected were due to environment and not inherited. Fig. 1 shows the result of selection for number of bristles on the scutellum (part of the thorax) of the fly *Drosophila melanogaster*. After twenty generations no further progress was made during another forty generations. And as a nearly pure line had been obtained selection in the opposite direction was almost equally ineffective.

The pure line theory has recently been severely criticised by Pearson (1930). But the experiments on which his criticism is based are largely on vegetative reproduction in Protozoa, where selection

TABLE I[1]

Year	No. of Beans	Mean Weight of Parents			Mean Weight of Offspring		
		Light	Heavy	Difference	Parent Light	Parent Heavy	Difference
1902	145	600	700	100	631.5±10.2	648.5±7.6	+17.0±12.7
1903	252	550	800	250	751.9±10.1	708.8±8.9	−43.1±13.5
1904	711	500	870	370	545.9±4.4	566.8±3.6	+20.9±5.7
1905	654	430	730	300	635.4±5.6	636.4±4.1	+0.9±6.9
1906	384	460	840	380	743.8±8.1	730.0±7.2	−13.8±10.8
1907	379	560	810	250	690.7±7.9	676.6±7.5	−14.1±10.9

[1] After Johannsen 1909. Weights in mgms.

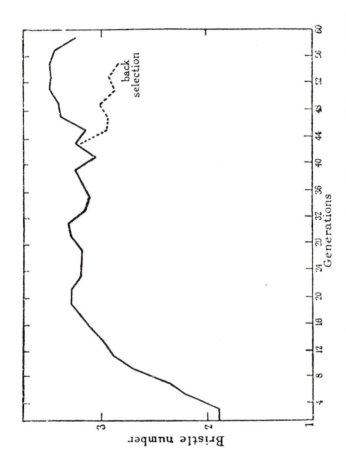

Figure 1. Effect of selection for high (full line) and subsequently low (dotted line) bristle number in *Drosophila melanogaster* (after Payne, 1921).

has a slight effect within the progeny of one individual. Such a race, however, is not a pure line in the sense in which that phrase is applied in the genetics of multicellular organisms. Moreover, Pearson believes that a pure line, though homogeneous, tends to deviate progressively from the original type. This is not borne out by experience. Thus fifty years of self-fertilisation have led to no progressive changes in many of the de Vilmorin wheats.

Such was the position of the selection theory ten years ago. It was shown that Darwin had been wrong in supposing that variations due to environment were inheritable. Selection merely picked out the best available line from a given population, and would not, as Darwin had believed, give rise to an unlimited amount of change. A number of biologists consequently proclaimed their belief that natural selection could not account for evolution. But no satisfactory alternative was forthcoming. The Lamarckian principle had been even more completely disproved than the Darwinian. On the top of Johannsen's work came such experiments as those of Payne (1911), who grew *Drosophila* for sixty-nine generations in darkness and found that neither the size of their eyes nor their tendency to move towards the light had been altered. Lamarck had believed that just as organs of an individual atrophy from disuse, this atrophy may be transmitted to their descendants. But wherever sufficiently careful experiments have been done, this has been shown not to occur. We shall have, later on, to consider some other theories of evolution which have been put forward. But I propose to anticipate my future argument to the extent of stating my belief that, in spite of the above criticisms, which are all perfectly valid, natural selection is an important cause of evolution.

While the geneticists were disproving many of Darwin's ideas, the palaeontologists were determining the actual historical facts of evolution. Where the data were adequate they were able to verify the law of succession, first explicitly given by Darwin's colleague Wallace. "Every species has come into existence coincident, both in time and space, with a pre-existing closely allied species." This would clearly be true on *any* theory of evolution, and probably false on a theory of numerous successive special creations. The evidence is most conclusive where we have records of marine life in fairly uniform conditions over many million years, as in the Welsh mountains and the English chalk. It is naturally less satisfactory for land animals, where the geological record is never quite continuous over very long periods.

But palaeontology has done far more than that. It has actually enabled us to follow the course of evolution in great detail, particularly in the case of marine organisms. The cases which have attracted most attention are those which clearly demonstrate slow and continuous evolution. Thus in a number of cases a species of mollusc producing a shell sufficiently like that of the common oyster to be placed with it in the same genus *Ostrea*, has gradually developed into something more like a cockle. The final forms are placed in the genus *Gryphaea*, which has been extinct since the age of the chalk. Now the process was gradual, If we collect a number of shells of the evolving species at any level, we find a certain type commonest, and others, more and less coiled than the type, somewhat rare. But at any level we could pick out from the population a few individuals representing the most frequent type a hundred thousand years earlier or later. Evolution in such cases has clearly been a very slow and almost (if not quite) continuous process, exactly as Darwin had predicted.

We must remember, however, that the organisms studied in this way are far from representative. They are in general the most successful members of animal associations living in very constant marine or lacustrine environments. We have not got similar data for land species, because the record, for obvious reasons, is not continuous over very long periods. Nor do we posses them for the rarer forms. We shall see later that perhaps dominant species in a uniform environment are the least likely to undergo sudden change to a new type.

Even in the record of the dominant marine forms there are breaks which suggest that some more sudden process was at work. Such is the break in the ammonite series which occurs in the Rhaetic. Along with the old Triassic types which they were to displace, new forms appear which were the ancestors of all later Ammonites. The palaeontologist can always postulate a slow evolution in some area hitherto unexplored geologically, followed by migration into known areas. But until a continuous series is discovered sceptics may well ask whether the gap, which is not a very vast one, was not bridged by a discontinuous process.

Further observation of these marine races showing slow continuous evolution displayed an extraordinary group of phenomena which are not obviously explicable on any theory of evolution whatever. Characters appear to go on developing past their point of maximum utility. Thus the coiling of the *Gryphaea* shells went on

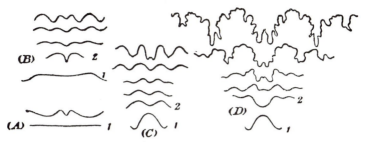

Figure 2. Development of suture lines in four Ammonites: (*A*) Anarcestes, (*B*) Tornoceras, (*C*) Glyphioceras, (*D*) Dactylioceras; 1. First septum, 2. Second septum. (*From Swinnerton's "Outlines of Palaeontology," Edward Arnold & Co.*)

until it must have been very difficult for them to open at all, and impossible to open widely. This state of affairs occurred several times, and always portended the extinction of the race. The same thing sometimes happened in land animals. Thus in the Titanotheria (large Oligocene hoofed mammals) gigantic size and horn development were the prelude to extinction in a number of separate lines of descent. One is left with the impression that the evolutionary process somehow acquired a momentum which took it past the point at which it would have ceased on a basis of utility.

But sometimes another process occurred, which has been particularly studied in the Ammonites. These animals, which in a general way resembled cuttlefish, made spiral shells with many chambers, but only lived in the last of them, the other being presumably filled with water or gas. The inner chambers were made by the young animals, the latter by the adult. So we can contrast the shell-making activity of the same animal at different ages. We then find that the earlier chambers often resemble those produced by the adults of ancestral forms some millions of years earlier. The phenomenon can be especially well studied in the suture lines between different chambers. The correspondence is not exact, and often new features appear in the earlier stages which were not present in any ancestors. Fig. 2 shows the development of suture lines in four Ammonites belonging to the early Devonian, late Devonian, middle Carboniferous, and late Lias respectively, a span of some two hundred million years. This is quite analogous to the phenomenon of partial recapitulation seen in the early development of such forms as man. An early human embryo has rudimentary gill-slits and a tail. Later on it develops a thick coat of hair which is shed before birth. Of course the gill-slits and tail are unlike those of any adult animal, and it has special organs such as the umbilical cord, which are not and never were found in adults. But many of its features recapitulate those of its adult ancestors.

All this can be explained on Darwinian lines. The less a new adult character interfers with normal development the more likely it is to be a success. When, however, it has been fixed in the adult stage the complicated developmental process may well be slowly modified so that the advantages of the new character appear earlier and earlier in the life-cycle and its appearance is less and less abrupt. This process is, however, likely to be very slow.

So far so good, but in the later stages of Ammonite history a much more surprising phenomenon occurred. A number of different lineages began to alter in the opposite direction. Features appeared which had not been seen for a hundred million years, but which strongly resembled those of the earliest known Ammonites. The suture-line became simplified, and the shell uncoiled. Sometimes the primitive features seem to have been present right through the animal's life history. In other lines of descent (e.g. *Baculites*) the shell was at first coiled, but in the fully adult animal it was straightened out. This reversion to primitive types was always the prelude to extinction. It happened on a large scale in the late Trias, when most of the great Ammonite groups died out. Then there was a brilliant renaissance during the Liassic period, one of the older groups giving rise to many new types. But an epoch of archaism set in once more in the Cretaceous, and at the end of that period the last Ammonite died. The closing stages of Ammonite evolution were marked, not only by retrogression, but by the appearance of new shell types, with "hairpin bends" as in *Hamites*, or an asymmetrical snail-like spire as in *Turritelites*. These bizarre forms, however, were only temporarily successful. After about 400 million years of life the Ammonites became extinct.

The account here given is that due to Hyatt and Wurtemberger, and is, I think, accepted by most palaeontologists. However, Spath's (1926) views on Ammonite lineages, which are easier to reconcile with Darwinism, command much support. I am not competent to judge between them, but wish to state the anti-Darwinian position as fairly as possible.

Now this process of "racial senescence" was not peculiar to the Ammonites, although it can best be studied in them, owing to the fact that their early stages are preserved. It seems to have occurred also in the Graptolites, Foraminifera, and other groups. The preservation, in the adult stage, of what were embryonic characters in the ancestor is called neoteny. It was probably one of the processes concerned in the retrogression of the Ammonites. When the embryonic stage whose features persist in the adult was itself primitive, neoteny clearly leads to a partial reversal of the evolutionary process. Often, however (to some extent even with the Ammonites), this is not the case. In the course of evolution features appear in embryonic life which do not correspond to anything in the ancestral series. Such is the mammalian placenta (originally developed as a respiratory organ within the egg, but unrepresented in fish). The appearance of novel embryonic features is called caenogenesis. When neoteny supervenes on caenogenesis, although certain features of the ancestral adult are lost, new characters appear which have not previously been seen in adult ancestors, and thus important evolutionary novelties come into being. This combination, *i.e.* neoteny supervening on caenogenesis, seems to have occurred in human evolution. Man is far more like a young gorilla or chimpanzee than an adult, and perhaps even more like a foetal one. In human evolution some characters of recent ancestors, such as the hairy coat, have been thrust back to pre-natal life. But as regards the general shape, and especially the head form, it is more nearly true that the final stages of individual development have been left out. Bolk has called the process leading to man foetalisation. I shall have to discuss it again later.

The story of the Ammonites is not very easy to reconcile with evolution by natural selection. And while acceleration of development, *i.e.* pushing back of adult characters into early life, might be explained on a neo-Lamarckian view as due to the cumulative action of something like racial memory, the reverse process would involve a progressive racial forgetting of certain tendencies. Nor are the facts any more consonant with the view that evolution represents the working out of a purpose, and is intelligently directed. On numerous occasions related species have gone through very similar changes as a prelude to extinction. We should have to suppose the directing mind intelligent enough to design new types of organism (perhaps only a biochemist can form an adequate idea of the difficulties of doing this), but not intelligent enough to learn from its own mistakes. For the above reasons many palaeontologists to-day confine themselves to stating the facts of evolution, and laying down general laws which they obey, rather than attempting to discover the causes underlying those laws. Apart from any hypothesis, it seems likely that, for example, the birds and the gastropod molluscs are now at or near their maximum of complexity, success, and variability, the mammals perhaps slightly past it, the reptiles very definitely so, and the amphibians still more markedly on the down-grade.

Meanwhile, however, Darwinism was attacked from quite a different angle by naturalists and some geneticists. As an example of the criticism of an extremely competent student of wild life I should like to cite Willis' "Age and Area" (1922), a book packed with facts which any theory of evolution will ultimately have to incorporate. The fact that it offers no theory as to the causes of the evolutionary process may explain the (to my mind) entirely unmerited neglect of the data presented in it. Willis first produced strong evidence that a number of rare plant species of restricted habitat were new, as opposed to relicts of species now dying out. For example, of the 809 species of flowering plants found only in Ceylon, about 100 were confined to the tops of single mountains, and 200 to very restricted areas. But the areas of the 228 moderately rare species overlapped in every conceivable way, like coins thrown down at random. All this is quite intelligible if the very rare species are mostly new, the moderately rare a little older, and so on. It is unintelligible on the hypothesis that these species are old ones in the course of dying out. If so one would have expected that a very rare species would frequently be found in two or three isolated spots wide apart. This was sometimes so, but very uncommonly. A large amount of other evidence agreed with the same hypothesis.

Clearly if it is true we can study a number of newly born species. When we do so we discover that they differ sharply from the surrounding species. Now I suspect that some of Dr. Willis's rare species are after all dying relicts, and some few are mere varieties due to the action of a single gene with many different effects. But I have far too much respect for his ability as a taxonomist to suppose that this is often the case. When the difference between species are analysed genetically they usually turn out, as we shall see later, to be of a more compex character than those between varieties. The species *Coleus elongatus*, which consists of about a score of plants on the top of one mountain, differs from the widely spread *Coleus barbatus* found alongside it, in respect of fourteen different characters. There are no intermediates. Unfortunately their genetics are unknown. Willis, then, believes that the birth of a new species from an old is often a sudden process. The new species must, of course, justify its existence by surviving, but natural selection, on this view, does nothing to make the new species. It merely decides which of a large number of new species formed by mutation will survive. In the case of flowering plants Willis estimates the number of new species starting on a successful career at about two per century. For that reason it is intelligible that till recently the process had not been observed, though many varieties had originated under close observation.

From his studies of the genetics of *Oenothera*, de Vries (1904) was led to the conclusion that new species originate abruptly. Other geneticists took refuge in agnosticism on the ground that nothing comparable to a specific difference had even arisen in cultivation. Thus Bateson (1928) said, "In dim outline evolution is evident enough. . . . But that particular and essential bit of the theory of evolution which is concerned with the origin and nature of *species* remains utterly mysterious. . . . The production of an indubitably sterile hybrid from completely fertile parents which have ariven under critical observations from a common parent is the event for which we wait. Until this event is witnessed, our knowledge of evolution is incomplete in a vital aspect." I may add that this event has since occurred.

It will be seen that the evidence from palaeontology and from modern rare species is contradictory. This is natural, because species rare in their own day are in all probability absent from the geological record. Also we have no really satisfactory evidences of perfectly continuous evolution in plants, where the evidence of abrupt species formation is strongest.

To sum up the situation so far, we may say that the criticism of Darwinism has been so thorough-going that a few biologists and many laymen regard it as more or less exploded. At least we may claim to have cleared the ground for an impartial survey of the facts. In the remaining chapters I shall try to answer the following questions: What is the nature of heritable differences within a species? Are the differences between species of the same or of a different character? Does selection really occur in nature? If so, will it account for the formation of species? Must we allow for other causes of evolutionary change? And, finally, when we have surveyed the process of evolution we shall have to ask what judgment we can make about it. Is it good or bad, beautiful or ugly, directed or undirected? These are largely value-judgements, and are thus not scientific. But it is the answer to them which makes evolution interesting to the ordinary educated man and woman. In making them I can, of course, claim no special standing. I can write of natural selection with authority because I am one of the three people who know most about its mathematical theory. But many of my readers know enough about evolution to justify them in passing value judgements upon it which may be different from, and even wholly opposed to, my own.

CHAPTER II

Variation Within a Species

"Varieties, as we shall see, may justly be called incipient species."—
Darwin.

The individuals belonging to a species differ to a greater or less extent. We can divide the causes of variation into those which operated before and during the life of the individual. We take that life as beginning with the fusion of the nuclei of the gametes which formed it, namely, the egg and the spermatozoon in most animals, the ovule and pollen grain in higher plants. (The organism produced by the fusion of the gametes is called a zygote.) In many plants and a few animals we can study the effects of nurture, *i.e.* causes operating during the life of the individual, almost apart from those of nature, *i.e.* causes operating earlier. When a plant or animal can be propagated vegetatively, the vegetative progeny of a single individual resemble one another to an extraordinary degree, and are called a clone. Thus all the Cox's Orange Pippins in the world are grafted from one seedling. The differences which exist between members of a clone are mainly due to environment and not to heredity. Thus Cox's Orange is a very different plant according as it is grafted on French Paradise, which gives a suburban garden bush, or Broad-leafed English Paradise, which gives an orchard tree, yet in a given environment it behaves in a predictable way. Even within a clone new types may appear (so-called bud-sports). These generally produce their like when vegetatively propagated. But with these exceptions, differences within a clone are not inherited. They are the best example of what is called fluctuating variability, due to differences of environment, not transmissible by inheritance, and therefore irrelevant for the problem of evolution.

You cannot propagate guinea-pigs by cuttings, but by many generations of inbreeding you can produce a line of guinea-pigs extraordinarily alike. After twenty or more generations of brother and sister mating there is no more resemblance between parent and offspring than between cousins. You do not abolish variation, and if you choose a piebald race it is easy to study it. The pattern is affected by environment, especially by the age of the mother, but these variations are not inherited. Wright, 1926).

Now attempts are constantly being made to prove that differences due to different environments are inherited. We shall see that this is true in a few cases. But in the vast majority of the experiments (as far as I know, in all but one) on which the neo-Lamarckian case is founded, no attempt has been made to establish a pure line to start with. It is therefore impossible to say whether the variations which are observed are not, at least in part, due to internal causes—that is to say, nature rather than nurture—and therefore determined and inherited according to the laws of ordinary genetics.

Table I (p. 9) gives some idea of the immense differences which may exist within a pure line, and the fact that they are not to any appreciable extent inherited. It is commonly supposed that the case against Lamarckism is largely based on the *a priori* arguments brought forward by Weismann. Weismann pointed out that in a higher animal such as man or guinea-pig the germ-plasm, which is to give rise in the next generation, is segregated at an early stage, and largely independent of the rest of the body. We now know further, what Weismann did not fully realise, that its chemical and physical environment is kept extraordinarily constant in a higher animal. Whereas in an insect the germ cells are at varying temperatures, which may affect their genetical behaviour, and in some lower animals at least the chemical composition of the blood varies a good deal with the external environment, this is not the case among us higher animals. Moreover, our gonads will function, at least for a considerable time, without connection with the nervous system. To quote the somewhat over-emphatic words of Claude Bernard (1878), "All the vital mechanisms, varied as they are, have only one object, that of preserving constant the conditions of life in the internal environment." He might have added "and thus to prevent the germinal transmission of acquired characters."

But in plants there is no such early segregation of the germ cells. They are formed like any other cells from the undifferentiated tissue of the growing points. In plants Weismann's *a priori* argument is worth nothing. We might expect to find evidence for Lamarckian effects among them. With one possible exception, to be noted later, we do not.

Let us turn to the facts concerning inheritable variation. Naturally we know most about variations of characters which are not readily affected by the environment, or which are at least stable in such a relatively constant environment as that of a breeding-pen or greenhouse, and it is mainly with these that I shall deal. A very rough classification divides these variations into six classes according to the mode in which they are genetically determined.

1. The simplest case is that of a character due to an extra-nuclear factor, or plasmon, as Wettstein (1928) calls it. There are two races of *Primula sinensis*,[1] one with green, one with yellowish leaves. No matter what pollen we use, the seeds of the green plant will produce only green seedlings, of the yellow only yellowish seedlings. We can use pollen from the yellowish plant on green plants for many generations, but never get a trace of the character carried over. The reason is quite simple. The leaf colour is due to chloroplasts in the cells. All the chloroplasts in the egg are contributed by the mother. The father contributes none to the pollen grain. Other characters than leaf colour can be affected by plasmons. In flax (Gairdner, 1929) the sexuality of the plant depends on a balance between the plasma outside the nucleus, contributed by the mother only, and the genes in the nucleus contributed by both parents. Upset this balance, and you get plants with no pollen. It is very likely that some of the remarkable results of Goldschmidt (1920) on sexuality in moths are due to cases of the same kind. Unfortunately, however, in this case one of the chromosomes is contributed by the mother only, so we cannot be sure of the importance of extra-nuclear factors.

2. The next simplest case is that of a character determined by a single Mendelian factor or gene. I am not going to give a full exposition of Mendelism, but just to recall some of its essential features. Fig. 3 shows a normal individual of *Primula sinensis*, and one with the longer type of leaf called "fern." If we start with pure lines of normal and fern leaf, and cross them either way, we get normal-leaved hybrids. Crossed with normal they give normals, with fern leaf half normal and half fern. Half the pollen grains and half the eggs of the hybrids carry something which we call Y, whose presence in a plant causes the leaves to be, roughly speaking, round instead of long. The other half carry something called y. In this case we cannot pick out the pollen grains carrying Y from those carrying y, but this is possible in some similar cases. Normal leaved plants

---

[1] For a full account of the genetics of this plant, see de Winton and Haldane (1932).

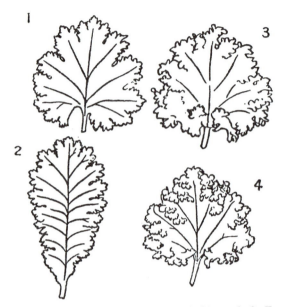

Figure 3. Types of leaf in *Primula sinensis*. 1, Normal. 2, Fern, *yy*. 3, Slightly crimped, f^sf^s. 4, Strongly crimped, f^lf^l The last two are due to genes allelomorphic with one another and with the gene for flat leaf.

may be of composition YY like the pure line of normals, or Yy like the hybrids. Fern-leaved plants are always *yy*. Y and *y* are called genes. Each cell of an adult plant or animal generally contains two of each kind. If they are alike, as in YY and *yy*, the organism is called a homozygote, and if unlike a heterozygote. Y and *y* count for this purpose as one kind, for *y* is really a modification of Y. Each gamete contains one gene only of each kind. A pair of genes related like Y and *y* are called allelomorphs.

In this particular case YY and Yy are indistinguishable to the eye. In other words, Y is dominant, and *y* recessive. There is a good case of incomplete dominance in the same plant. DD has white flowers with a pink flushed centre, *dd* has red flowers, D*d* is intermediate. Sometimes we only know two modifications of the same gene, sometimes quite a number. They are then called multiple allelomorphs, and generally affect the same character to different degrees. The same figure shows, also in *Primula sinensis*, a normal leaf and two types of crimped leaf caused by the genes allelomorphic with normality. You may have as many as eleven multiple allelomorphs. They obey a simple rule. You cannot get more than one gene of the series into a gamete, nor more than two into a zygote, *i.e.* an adult organism. A set of multiple allelomorphic genes affect the same organ or character in different degrees.

By a study of organisms with too many or too few chromosomes it has become clear that any given gene goes with a certain chromosome.[1] If there are three chromosomes of a kind in a zygote, it contains three of the genes in question, and so on. Sometimes we can say whereabouts in a chromosome a given gene is to be found. Occasionally we can see this directly. You are all familiar with double stocks. They are sterile, but some races of single stocks produce seed of which a little over half give double plants. Cross ordinary singles with their pollen, and all the hybrids behave as ordinary heteroxygotes for the recessive character of doubleness. But if the ever-sporting race were heterozygotes we should expect only half their pollen to carry the recessive gene *s* for doubleness. What has happened to the pollen carrying the dominant gene S for singleness? It will not germinate. When this was discovered, and indeed before Snow and Waddington (1929) showed that it would not germinate, this was put down to its lack of a gene P needed for proper germination. P is somewhere in a little knob or trabant on the end of one of the chromosomes, and the double-throwing plants have only one P and only one knob (Philp and Huskins, 1931).

The main work on the location of genes has been done by Morgan and his colleagues in America (Morgan, 1926). They have shown by a study of the way in which they hang together in the offspring of animals heterozygous for several genes at a time that the genes are arranged in a row along the chromosome in a definite order of which a map can be made. Fig. 4 is a map of the four chromosomes of each of which there is a pair in the female of the fly *Drosophila melanogaster*. The map distances represent probabilities of interchange between the genes. They are not strictly proportional to their real distances, but the order is correct. When two genes are near together on the map, this means that if they have gone into a female zygote together they will probably come out in

---

[1] It will be remembered that the chromosomes are small bodies which are visible in a nucleus while it is dividing. They can be seen in living material, but they are best shown up by a number of stains. Hence their name. Their number and shapes are generally very constant in a given species, except that the two sexes may differ. Occasionally, however, individuals are found with too many or too few chromosomes.

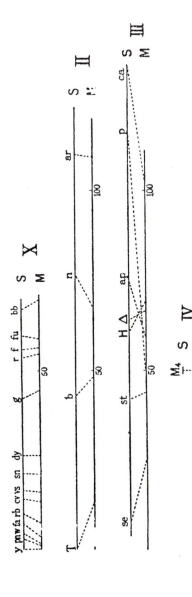

Figure 4. Maps of the chromosomes in *Drosophila melanogaster* (M) and *Drosophila simulans* (F) (after Sturtevant, 1929). Only genes which have mutated in both species are shown. The abnormal characters caused by them, reading from left to right, are as follows: Yellow body, Prune eye, White eye, Facet (rough) eye, Ruby eye, Cross-veinless wing, Singed wing, Dusky wing, Garnet eye, Rudimentary wing, Forked bristles, Fused wing veins, Bobbed bristles, Truncate wing, Black body, Nick wing, Arc wing, Sepia eye, Straw-coloured body, Peach eye, Aristapedia (legs in place of antennae), Delta wing veins, Hairless, Claret eye, Minute bristles.

the same gamete. The rearrangement is due to an exchange of parts between a pair of chromosomes, one derived from each parent. The nearer together two genes are in the chromosome, the less likely is an exchange which will separate them. If the distance between the genes is one unit, this means that the probability of their parting company, if they have gone in together, is one-hundredth.

3. Commonly we find that two races differ by several genes. If these genes affect quite different characters, *e.g.* hair length, hair colour, and fat colour, there is no difficulty in distinguishing them. If they affect the same character, *e.g.* body weight, the problem is much more serious. In *Primula sinensis* we know of eight genes which may affect stem colour. One is incompletely dominant, so there are at least 384 possible genetically different types of plant with (on the average) different stem colours. If we had members of all they would form a nearly continuous series. In certain crosses we have got forty-eight of the stem colours, and the series from green to deep purple did seem to be continuous. My colleague, Miss de Winton, who is a good enough geneticist to take Mendelism seriously, accepted this challenge, and we can now isolate the various stages in the series. This has been possible because all but one of the genes concerned have some marked effect other than that on stem colour. Most of them affect the colour of the petals, one that of the stigma, and so on. This enables us to label the individual genes, so to say, and determine their responsibilities.

No procedure of this kind is as yet possible in the case of such an apparently continuously variable character as the height of men or the weight of rabbits. Ultimately it may well be found that of the genes influencing human height some act through the thyroid gland, others through the pituitary, others through the gonads in delaying maturity, others again more directly on the bones, and so on. That is mere speculation. At present we cannot even prove conclusively that such continuously varying characters are due to genes at all. But we can render it extremely plausible. If we cross two fairly constant races, say Hamburgh fowls with a mean weight of 1300 grams[1] and Sebright bantams with a mean weight of 750, the hybrids are intermediate and fairly constant, just as when we cross green-stemmed and dark purple-stemmed Primulas. But on crossing these hybrids we get a wild outburst of variation, and in later

---

[1] Weights of cocks at thirty-five weeks. The reference is to Punnett and Bailey's (1914) experiments.

generations birds are obtained lighter than the bantams and heavier than the large parent, *e.g.* cocks weighing 1700 grams. If there are n dominant genes we have 2^n types, if the genes are not dominant 3^n. Thus ten genes would be enough to give 3^{10}, or 59,049 different types, in other words a range of variation which is for all practical purposes continuous. Qualitatively the inheritance of rabbit weight is as would be expected if it is mainly determined by multiple genes.

In man, where far more evidence is available, the agreement is quantitative. Pearson and his colleagues studied the inheritance of stature and other characters which appear to vary continuously, and obtained very definite results, not obviously explicable on any theory. They did not agree with Galton's law of ancestral heredity, nor with Mendelism as then understood. Recently Fisher (1918) has shown that these results agree exactly with the expectation if stature is determined by a large number of genes. As the data regarding human stature are far more precise and extensive than for similar characters in plants and animals a presumption is established that a similar explanation holds good in non-human cases.

4. Until recently it was thought that the order of the genes in a chromosome was something definite. It had long been known that when certain races of *Drosophila* are crossed the expected re-. arrangement of certain of the genes of the hybrid, which are carried by chromosome No. 3, does not take place. When maps are made of the genes in the third chromosomes, those which rearrange themselves are in the same order in both races. Those which will not interchange lie in a section of chromosome in which the order in the two races is opposite. It is as though the end section of the chromosome of one race, with all its genes, had been removed, and stuck on in the opposite order in the other. Several such cases have now been described as the result of treatment with X-rays in *Drosophila*.

There are other more serious kinds of aberration which involve an alteration in the arrangement of the genes, but not in their number or quality. For example, two chromosomes which are normally separate may be stuck together. In this case the genes in them, which are not normally linked, exhibit linkage.

Or a piece of one chromosome may be stuck on to another (quite a frequent result of X-ray treatment in *Drosophila*). Thus Dobzhansky (1929) obtained several races in which bits of the second or third chromosome had been stuck on to the fourth. The linkage results were in accordance with expectation.

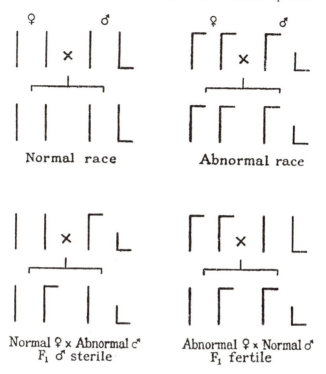

Figure 5. Diagrams of the sex chromosomes in two races of *Drosophila melanogaster* (after Stern, 1929). In the abnormal race a piece of the L-shaped Y chromosome has been attached to the X. This is necessary for fertility, and males which do not receive it from either parent are sterile.

From our point of view the most interesting case of this kind is one recorded by Stern (1929). It will be remembered that in many animals the female carries two X chromosomes, as they are called, the male an X and a Y. The Y contains very few genes, and is more or less of a dummy. A male Drosophila without it looks normal, but is sterile. For fertility this chromosome, or at least a large part of it, must be present, but at least part of it may be attached to the X chromosome without harm. Stern studied an aberrant but quite fertile race (obtained by X-raying the normal race) in which the long arm of the J-shaped Y had been stuck on to the X, and a fragment had been broken off the Y, so that its two arms were of equal length. The cytological conditions are shown in Fig. 5. It will

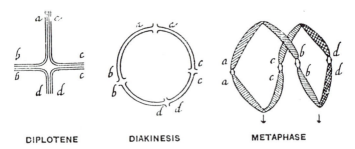

DIPLOTENE   **DIAKINESIS**   **METAPHASE**

Figure 6. Behaviour of chromosomes at meiosis in a species where segments have been interchanged between two of them. In the hybrid four chromosomes pair as shown. Unless they separate as shown, inviable gametes or zygotes are formed. Thus a gamete containing chromosomes *ab* and *ac* is useless because it lacks *d*. Such hybrids are therefore usually semisterile (after Darlington, 1929).

be seen that when the two races are crossed the females are always fertile. In one cross the males are fertile, in the other they are sterile, because they do not contain a complete Y chromosome. We shall see the importance of this when we come to consider inter-. specific hybrids.

Occasionally halves of two chromosomes are interchanged. Calling the original chromosomes AA′ and BB′ the new pair is AB′ and A′B. At the reduction division such chromosomes pair in a set of four, which often opens out into a ring (Fig. 6). The genetical effect of this is to produce linkage between genes which are normally in different chromosomes, and therefore unlinked. Hammarlund (1923) detected abnormal linkage of two genes in the pea, and Håkansson (1929) found that the plants giving it had a ring of four chromosomes. As some of the types of gamete formed by the breakup of such a ring are inviable, a certain amount of sterility is found in plants with such a chromosome ring. But the two types whose hybrid forms the rings may be fully fertile.

5. We now come to a group of cases in which some, but not all, of the genes are represented more or less than twice in the aberrant type of individuals. This is the cause of sex-differentiation in most animals and some plants. In female mammals there are two X chromosomes, in the male an X and a Y. The Y contains very few genes, and perhaps in some cases none. From a feministic point of view maleness may be considered an aberration. Generally the

male is the heterogametic sex, with an unequal pair of chromosomes, or rarely with an X and no Y. But in birds, lepidoptera, and some fish, the male is homogametic and the female heterogametic. In plants it is not uncommon to find one chromosome represented three times. This usually causes quite noticeable morphological changes, generally, if not always, more marked than when there are three of all the chromosomes. In the latter case the number of genes of all sorts is increased equally, in the former the balance is upset. Such trisomic plants, as they are called, may be quite vigorous. But generally the pollen grains carrying the extra chromosome are less viable than the normal type, and often they do not function at all, at least in competition with the normal.

Datura Stramonium, the American Jimson Weed, is particularly prone to this type of aberration. It has twelve pairs of chromosomes, and also twelve easily distinguishable trisomic forms, in each of which one chromosome is represented thrice. There are further types in which the extra chromosome is composed of pieces of two of the normal ones. Blakeslee, Belling, and their colleagues (1928) have made an exhaustive study of this phenomenon in *Datura*, but it not uncommon elsewhere. Thus Darlington (1928) found that all the cultivated types of sweet cherry (except the "Dukes," which are tetraploids) have from one to three more chromosomes than the wild species. For the reason given above, such plants do not breed true from seed, but as they are reproduced by grafting this does not prevent their being useful in cultivation. In animals the presence of an extra chromosome generally produces a very unhealthy type, unless the chromosome in question is a very small one, or the Y sex chromosome, which, as it carries very few genes, is nearly a dummy.

6. The last type of heritable variation is due to the addition of one or more whole sets of chromosomes. If $2n$ is the normal number, an organism with $3n$ is called a triploid, with $4n$ a tetraploid, and so on, the general name for such plants being polyploids. Where all the sets of chromosomes are derived from the same species they are called autopolyploids.

Autotetraploid varieties are very common in cultivated plants. Not to go outside the Primulaceae, they are found in *Primula sinensis*, *P. obconica*, *P. malacoides*, and *Cyclamen persicum*. In these cases they have arisen in cultivation, and have been preserved for their large flowers. The origin, which has now twice been observed under absolutely critical conditions in *Primula sinensis*, is quite sudden. Tetraploidy generally leads to an increase in size and a diminution in fertility. The tetraploids are, however, fertile enough for commercial purposes except where, as in the tomato, the size of the fruit depends upon the number of seeds. But they are only fertile with one another. The cross with the diploid ($2n$) is generally either a failure, or gives rise to a sterile hybrid. Thus in *Primula sinensis* great effort has been devoted to crossing the diploid and tetraploid forms. When tetraploid pollen is put on a diploid the tubes generally grow up into the air instead of down the style. On one occasion, however, there is a credible record of a seedling being obtained from such a cross. The opposite cross occasionally gives a seed. Nine fertile seeds had been obtained from it up to 1929. Seven of these gave rise to triploid plants (with three sets of chromosomes). Their nuclei divide unequally, so they are very sterile. Two gave rise to tetraploids, having apparently been produced by unreduced pollen grains with a double set of chromosomes.

In the tomato it is possible to produce tetraploids at will. If we take a hundred tomato plants, cut them down, and cut back the new shoots until the plant is nearly exhausted, about 6 per cent will produce a branch with larger leaves, which is a tetraploid. The only account yet published is that of Jørgensen (1928). If the flowers on such branches are self-fertilised, the seedlings are also tetraploids, quite easily distinguished from the ordinary diploid. We have here the exception which, so to speak, proves the very general rule that the effects of injury are not inherited. They are inherited in this case because the injury has provoked a stable type of rearrangement of the nucleus. It would seem that the nuclear changes associated with ordinary reactions to the environment are reversible, while induced tetraploidy is not, except under very unusual circumstances. It is also noteworthy that in this case only those germ cells are affected which are actual descendants of injured cells. There is no effect whatever on the seedlings from branches which have not become tetraploid. Hence no support is offered to the view that the effects of injury, use, or disuse of a part might be carried over to the offspring in a higher animal, where the germ-cells are early separated from the somatic cells. In the tomato the diploid selfed yields about seventy fertile seeds per fruit, the tetraploid about twelve, but rather more fruits per plant than the diploid. The triploid, on the other hand, gives much less than one viable seed per plant per year, though about a dozen fertile seeds all told have been obtained from triploids self-fertilised or crossed with other triploids.

It is noteworthy that we have here a case of the origin, spontaneous or provoked, of a variety so different from the original type as either to refuse to hybridise with it, or to give sterile hybrids. Such behaviour was considered in the past to be the note of a specific difference. Huxley and Romanes lamented that it could not be produced artificially. To-day Catholic apologists, whom I sometimes read, because their arguments are at least coherent, still taunt us poor Darwinians with our failure, though R. P. Gregory's account of tetraploids and their origin and genetical behaviour, was published in 1914. I should like to take this opportunity of calling attention to the work of Gregory, whose name would by now be familiar if he had not died of influenza in 1918 at the height of his powers.

All hereditary differences which have been thoroughly investigated seem to fall into one or more of these six classes. A few, such as the "rogue" character in peas, remain mysterious, but are still under investigation.

How do these intraspecific differences originate? Partly no doubt by combinations of those which existed before. If we have races of white rose comb and black ordinary comb poultry, it is very easy to make the new combination of white normal and black rose. But this is not the whole story. New genes arise from time to time by a process called mutation. The majority of new genes are recessive to the wild type, and are probably wild type genes which have wholly or partly lost their activity. But some at least are dominant. As a dominant gene produces a visible effect on its first appearance, while a recessive must wait for a generation, we know more about the origin of dominants than of recessives, although they do not seem to differ in principle. In *Drosophila melanogaster* at a time (1925) when about fifteen million individuals had been bred from known parents and fairly carefully inspected, the principal gene determining eye colour had been observed to mutate twenty-five times, no other gene having mutated so often. This gives an upper limit of about 10^{-4} for the mutation frequency of a given gene per life-cycle if we allow for the fact that only about 1% of the flies are bred from. On the other hand, many lethal mutations occur in *Drosophila*, and some of these may be commoner, though this has little relevance for evolution. In maize, on the other hand, Stadler (1930) conducted an experiment involving the counting of a million and a half seeds. Seven out of eight of the normal genes mutated at least once, and one gene had a mutation frequency of about

4×10^{-4} per generation. Such high values are entirely exceptional. Gregory, de Winton, and Bateson have grown over 200,000 *Primula sinensis* under close observation. No mutation has occurred more than once, so far as is known, though about one visible mutation of one kind or another occurs in 20,000 plants. We can get some idea of the frequency of mutation in man by considering the frequency of rare and very disadvantageous genes such as that causing haemophilia (failure of the blood to clot). Here the rates of production by mutation and elimination by natural selection must about balance, and the probability of mutation of the normal gene works out at about 10^{-5} per life-cycle. One important point is that mutation is a sudden process. A single gene alters, and the alteration takes place at once and not by successive steps.

The fundamental importance of mutation for any account of evolution is clear. It enables us to escape from the impasse of the pure line. Selection within a pure line will only be ineffective until a mutation arises. Among a few million individuals a mutation of the desired type is not unlikely. Among a few thousand it is most improbable.

The rate of mutation can be enormously increased in two ways at least. Muller showed in 1927 that in *Drosophila* it was increased about 150 times by X-rays. β rays from radium are equally effective. It was natural to attribute normal mutation to high frequency radiation or rapidly moving electrons from potassium or other radio-active bodies, or other sources. Muller (1930) has shown that this is very improbable. Muller's work has been repeated, with similar results, both on *Drosophila* and other animals and plants. In 1929 Goldschmidt showed that mutation could be induced in *Drosophila* by heating the eggs to such a degree as to kill most of them. The mutations obtained were not, like Muller's, at random, but there was a specially large yield of two new types, a dark body and an abnormally veined wing. Jollos (1930) has confirmed Goldschmidt, Rokizky (1930) has partially done so. Ssidorov, Ferri, and Shapiro (1929) have failed to. Rokizky's work suggests that heat may produce an instability among the genes not culminating in mutation until after some generations. Harrison (1928) reported the induction of melanism in several species of moths by feeding them with lead and manganese salts. The melanism, when it had once appeared, behaved in a Mendelian manner. As, however, Hughes (1931) has failed to repeat the work, though using much larger numbers of one species than Harrison, it seems unjustifiable to draw very far-reaching conclusions from this work. There is no doubt, however, that mutation rate varies with external conditions. A few abnormal genes are very mutable. Such are those responsible for flaked flowers in many plants, and for a few characters in *Drosophila*. They are recessive, but have a tendency to mutate back to the normal with a probability varying from about 0.3 to 10^{-4} per generation. In these cases the mutation frequency is undoubtedly influenced by other genes. There is some inconclusive evidence that this is also the case with the mutation of normal genes. But whether this influence is highly specific or general we do not know, though the latter seems more likely.

Some kinds of chromosomal aberrations are quite common. For example, in *Drosophila melanogaster* the two sex chromosomes go to the same pole about once in 2000 reduction divisions, thus producing zygotes with too many or too few chromosomes. X-rays will cause irregular nuclear divisions and breaking of chromosomes. Similar results are produced in nature by such plant parasites as mites (Kostoff and Kendal, 1929).

Such, then, are the main causes of variation within a species. It is important to realise that none of these were known to Darwin. Mendelian inheritance was only discovered in 1900. The other four causes have gradually been discovered since that date. But while some writers on evolution have considered Mendelism, they have paid very little attention to other modes of variation.

Before leaving this topic I should like to guard myself against certain suggestions. There is a tendency in some quarters to describe the phenomenon with which I have just dealt as "the mechanism of heredity," and to suppose that the introduction of atomism by Mendel has reduced genetics to biophysics. I do not think that this is so. We can, in principle at least, speak of the mechanism of segregation. But the things segregated, the genes, reproduce themselves or are copied at each cell division. And this process of reproduction cannot at present be explained in physico-chemical terms, whatever may be possible in the future.[1] But it is a common-place of biology. The genes are biological atoms, just as the struggling individuals of Darwinism are regarded essentially as organisms, not machines. It is at present irrelevant to genetics whether life is or is not ultimately explicable in terms of physics and chemistry. Hence it is irrelevant to the general argument of this

---

[1] For a step in this direction see Haldane (1932c).

book, which is based on the facts of genetics. Genetics can give us an explanation of why two fairly similar organisms, say a black and a white cat, are different. It can give us much less information as to why they are alike. In the same way a complete theory of evolution might tell us how the various different species had originated from common ancestors. But it would give us little direct information concerning the nature of life.

CHAPTER III

The Genetical Analysis of Interspecific Differences

"All flesh is not the same flesh; but there is one kind of flesh of men, another flesh of beasts, another of fishes, and another of birds."—*St Paul* (I Cor. XV)

It is unfortunately impossible to give a satisfactory definition of the term "species" as used in zoology and botany. In many cases the species may be defined as a group of organisms which can breed together without loss of fertility in the first or subsequent generations. But this will not apply to organisms which do not reproduce sexually. You cannot cross two dandelions, but it would be very unsatisfactory for that reason to divide up the species *Taraxacum officinale* into some thousands of different species. No doubt systematists have sometimes based specific rank on trivial differences of morphology, and at other times have included within one species organisms which will not breed freely together. Nevertheless genetical work usually supports the opinions of systematists as to the more fundamental nature of specific than varietal diversity. In what follows I shall mainly deal with the analysis of differences between species which are distinct on any reasonable criterion. (For a discussion see Robson (1928).)

If Darwin was correct we should expect constantly to find difficulties in separating recently formed species. On the view here adopted specific differences are sometimes clear-cut from the first. The species problem is quite typical of the problems of science. We are compelled to investigate before we know what we are investigating, and as our knowledge increases we must continually restate our questions. For this reason, although some of the observed results recorded in this chapter are clear enough, the conclusions drawn from them will certainly need restatement in the future.

Species are usually defined by morphological differences, occasionally by chemical ones (*e.g.* of flower or feather colour), more rarely by differences of physiology, as in the case of yeasts or bacteria differentiated by their capacities for fermenting different substances. We are only at the very beginning of an analysis of the causes of these differences. If we want to analyse the difference between two varieties of one species, say a Manx tabby short-haired cat and a tailed blue long-haired cat, it is often sufficient to cross them and mate the offspring together or with the parents. In the case cited we should find the differences due to four dominant genes, causing short tail, short hair, banded hairs in certain areas, and dense pigment respectively. Sometimes we can do this with species, but rarely is the analysis complete. Often one of two things happens. The species will not hybridise, or else the hybrids are sterile like mules, or of one sex only, like fowl-pheasant hybrids, which are all cocks and sterile to boot. Even then, as we shall see, the geneticist, aided by the cytologist, can tell us a great deal. But before we pass on to the results of such analyses, we must consider two further topics—namely, comparative genetics and allopolyploidy. When we have a number of related species (or groups of species which are fertile *inter se*) and compare their genetics, we generally find a striking parallelism. To take two examples familiar to animal breeders, in most domesticated and many wild species we find a variety with white hair and pinkish eyes, the so-called albino. This pretty well always behaves as a simple Mendelian recessive to the normal form, the apparent exceptions being due to the fact that occasionally several other genes together may produce a pink-eyed white. Black-eyed and blue-eyed whites behave quite differently. Another extremely common recessive type is the long-haired, often called an "Angora."

In the case of the albino at least there can be very little doubt that the pink-eyed white in different species are due to inactivation of the same gene. I use the word "same" to denote homologous structure and similar function, as I might refer to the eye as the same organ, speaking of a rabbit and a mouse. Actually, however, the sameness may extend to molecular structure. The principle of homology between genes extends to a large number, as is clear from Table II. This is taken from a paper of mine published in 1927, but since publication a number of the gaps have been filled. For example, d^* was discovered in *Mus norvegicus* by Roberts in 1929. In the table + means that the new gene type is present in a wild

---

* *I.e.* the recessive allelomorph of the gene D present in the wild race.

form; but where several allelomorphic genes, *i.e.* modifications of the same gene, are present in wild types, the letter W is used. The

TABLE II

Gene	Effect	Mouse	Norway rat	Black rat	Deer-mouse	Cavy	Rabbit	Dog	Cat	Ferret
C	Normal	+	+	+	+	+	+	+	+	+
c^k	Slight dilution	−	−	−	−	D	D	−	−	−
c^d	Marked dilution	−	−	−	−	D	D	−	−	−
c^r	No yellow	D	D	−	−	D	D	D	D	−
c^b	Acromelanistic	D	−	−	−	D	D	?D	−	−
c^s	White	D	D	−	D	−	D	−	−	D
A^y	Yellow	D	−	−	−	−	−	?D	−	−
A^w	Light-bellied gray	W	+	+	+	−	+	+	+	−
A^g	Gray-bellied gray	+	−	W	−	+	−	−	D	−
A^r	Ticked-bellied gray	−	−	−	−	W	−	−	D	−
a^t	Black-and-tan	−	−	−	−	−	D	−	−	−
a	Black	D	D	D	−	D	D	D,?W	D	?+
E^d	Black	−	−	W	−	−	D	−	−	?+
E^s	Black	−	−	−	−	−	D	−	−	−
E	Normal	+	+	+	+	+	+	+	+	?+
e^p	Bicoloured	−	−	−	−	D	D	D	−	−
e	Yellow	?D	−	D	D	D	D	D	D	D
b	Cinnamon	D	−	−	−	D	D	D	−	−
r	Red-eyed yellow	−	D	D	D	−	−	−	−	−
p	Pink-eyed yellow	D	D	−	−	D	−	−	−	−
s_m	Salmon-eyed	−	−	−	−	D	−	−	−	−
i	Dilute	D	−	D	−	−	D	D	D	−
f	Yellow diluted	−	−	−	−	D	−	−	−	−
k	"Kodak"	−	−	−	−	D	−	−	−	−
h	Black slightly diluted	?D	−	−	−	−	D	−	−	−
D	Black intensified	?D	−	−	−	−	−	D	D	?+
u	Bicoloured	−	−	−	−	−	−	D	−	−
T^l	Lined tabby	−	−	−	−	−	−	−	+	−
T^s	Striped tabby	−	−	−	−	−	−	−	W	−
t^b	Blotched tabby	−	−	−	−	−	−	−	D	−
W	White	−	−	−	−	−	D	D	D	−
V	Piebald	D	−	−	−	−	D	D	?D	−
s_1	Piebald	D	D	−	−	D	D	D	−	−
s_2	Piebald	−	−	−	−	−	D	D	−	−
s_3	Piebald	−	−	−	−	−	D	−	−	−
s_4	White nose, feet or tail	D	−	−	D	−	D	D	−	−
R_e	Roan	−	−	−	−	D	?D	D	−	−
s_i	Silvered	?D	−	−	−	−	?D	−	−	−

letter D means that the new gene is present in a domesticated type, but only as a rarity, if at all, in wild nature. The parallelism is obvious. Similar results have been obtained by Vaviloff (1922) in cereals and other plants, though his genetical analysis is less complete. Some of Vaviloff's results are reproduced in Berg's "Nomogenesis." I shall use Table II as my text on several occasions, but equally good examples could be given from plants and insects.

When we cross two species the hybrids may be sterile. But the results are still sometimes of great genetical interest, owing to the production of what is called an allopolyploid. An example taken from the work of Newton and Pellew (1929) will make my meaning clear. The frontispiece shows two species of *Primula*, *Primula floribunda* from the Himalayas, and *P. verticillata* from Southern Arabia. It will be seen that they differ in several respects; in particular, *P. verticillata* is covered with meal. They can be crossed with some difficulty. The hybrid is vigorous, but, like the mule, almost absolutely sterile, though, unlike the mule, it can be propagated by cuttings. The sterility occurs for a somewhat different reason. Each cell of the mule contains a set of horse chromosomes and a set of donkey chromosomes, which co-operate to produce the characteristics of the mule. When the time comes to halve the number of chromosomes so as to produce gametes the machinery breaks down, and monstrous spermatozoa with too many, or defective with too few, chromosomes are produced. Things are not quite so bad in the hybrid *Primula*. The chromosome number of 18, 9 from each parent, is reduced normally, but the gametes almost all die, presumably because for viability a *floribunda* or a *verticillata* set is needed, and the chance of getting such a set is only 1 in 2^9, or 512. This number is really too small, because exchanges probably occur between the chromosomes of different species. Now when these sterile hybrids are grown from cuttings a surprising thing occasionally happens. A shoot is produced with somewhat larger leaves and flowers, which are quite fertile, and their seeds yield the well-known horticultural hybrid, *Primula kewensis*. The increased size is due to a doubling of the chromosome number, which is now 36; the fertility to the fact that each chromosome can find a proper mate, a *floribunda* chromosome pairing with another *floribunda* chromosome, and one going into each of the two gametes formed on reduction. Thus each gamete gets one complete set of *floribunda* and one complete set of *verticillata* chromosomes.

In consequence all the gametes are alike and the hybrid breeds true. This is, as a matter of fact, an exaggeration. A small amount of interchange takes place between *floribunda* and *verticillata* chromosomes, and *Primula kewensis* is variable for some characters, such as mealiness. It is also more liable than most species to drop a chromosome. But the aberrant forms are sufficiently rare to make the new plant a horticultural success. It is, as a matter of fact, much better adapted to horticultural requirements than either parent. Polyploids of this kind, which contain chromosomes from two different species, are called allopolyploids. They are quite common among plants, but although a case bordering on allopolyploidy has been reported in moths by Federley (1913) it is clear that among animals allopolyploidy is rare if it occurs at all.

With these prolegomena we turn to the main results of species crossing. In the simplest case the two species behave like varieties of the same species differing by several genes. Thus Chittenden (1928) investigated the results of crossing species within the Vernales section of the genus *Primula*, which includes the primrose, cowslip and the purple Caucasian *Primula Juliae*. The primrose *Primula acaulis* has yellow pigments in plastids, but no anthocyanin in its sap. *Primula Juliae* has sap pigment but no plastid pigment. The hybrid has both, and each is due to one gene. Thus when the hybrid is back-crossed to *acaulis* Chittenden got 130 with and 115 without anthocyanin (or nearly equality). Hence on selfing the hybrid we should expect to get one double recessive, lacking both sap and plastid pigment (*i.e.* a white) in 16. Chittenden got 4 out of 68. Other characters, *e.g.* the umbellate habit of the Bardfield Oxlip, *Primula elatior*, were shown to be due to single genes. Yet others, such as hairiness, were clearly due to several genes, but the discontinuous nature of the variation found in the second generation showed that the number of genes was not very large. In these cases the hybrids were rather sterile, so that Chittenden was unquestionably dealing with species crosses.

How are these genes determining interspecific differences related to those which determine the varieties on whose analysis Mendelism is based? In plants they seem to be of the same general character. In animals the investigation has not been carried so far, but the results are very interesting (for summary, see Haldane, 1927c). Turning to Table II, we notice that the gene A^g is present in the normal wild guinea-pig, *Cavia aperea* and its tame descendant, *C. porcellus*. It acts by making an anti-enzyme, of whose chemical properties we know a little (Koller, 1930), which inhibits the formation of black pigment. The inhibition is complete on the belly, which is yellow; on the back the hairs have alternate black and yellow bands, as in the wild rabbit. When this gene is inactivated as the result of mutation we get a black guinea-pig (*aa*). Wild colour is completely dominant to black. Now Detlefsen (1914), mated ordinary guinea-pigs to the wild species, *Cavia rufescens*, whose belly, instead of being yellow, has banded hairs like the back. The male hybrids are sterile, and it was necessary to backcross the partly fertile females for two generations with ordinary guinea-pigs before fertile males were obtained. When fertility was re-established some of the cavies had the *rufescens* type of coat. It was found that this character was recessive to the *porcellus* type, but dominant to black. It was in fact due to a gene which was a multiple allelomorph of A^g, differing from it less than the gene *a* of the blacks, or even a^t of the black-and-tan rabbit.

The colour difference between the geographical subspecies *alexandrinus* and *tectorum* of the "black" rat *Epimys rattus*, is due to a pair of alleiomorphs of the same kind. So is that between two races of mice which differ in habit rather than geographical distribution. Other rodent and carnivore species differ in the same way. Even the ferret and polecat, which have, perhaps erroneously, been placed in different genera, only differ as regards colour by a single gene. Of course the species and subspecies considered must differ by many other genes determining morphological, physiological and psychological characters. But as regards colour they differ less than the domestic races of one species. It seems probable, therefore, that in so far as interspecific differences can be analysed on Mendelian lines they are due to a number of small units of difference rather than a few large ones. It is at least quite certain that Mendelian gene differences, presumably due to mutation, have played a certain part in the origin of species.

In a small group of cases it can be shown that extra-nuclear factors, or plasmons, are partly responsible for interspecific differences. The clearest case is that demonstrated by Wettstein (1924, 1928) in the mosses *Funaria* and *Physcomitrella*. Reciprocal crosses give not only different hybrid sporogonia, but the spores from these give different series of haploid hybrids, and Wettstein showed clearly that this was due to extra-nuclear differences. Generally, however, reciprocal hybrids are fairly similar, showing that interspecific differences are mainly due to nuclear components.

Another case was discovered by W. C. F. Newton, and is still under investigation at Merton. The results were demonstrated to the Royal Society in 1926, but have not yet been published. When *Geranium striatum* and *G. Endressi* are crossed, the hybrid is more or less intermediate. Three genes of *striatum* and one of *Endressii* are dominant. The results of reciprocal crosses are indistinguishable in the first generation. But in F_2 from *striatum* × *Endressii*, though not from the reciprocal, rather less than a quarter of small-flowered, male-sterile plants appear. Such plants can be obtained in various ways. In every case it appears that to obtain a male-sterile plant a pair of recessive genes from *Endressii* must be present in cytoplasm derived from *striatum*. The parallelism with Gairdner's case in *Linum* is thus complete. There is a good deal of evidence as to interspecific differences in chloroplasts among plants, and much less conclusive evidence for cytoplasmic differences in animals.

While an analysis of the effect of several genes is often possible in the case of interspecific differences in colour or certain morphological features, it is more rarely so in the case of size. We usually find that the F_1 hybrid generation is fairly uniform, but when they are mated together or self-fertilised we get a wide range of variation, often including dwarf or otherwise abnormal types. This is what might be expected. Punnett and Bailey (1914) found that when the first cross between Bantam and Hamburgh fowls were crossed together, birds heavier than Hamburghs appeared in the second generation. The whole question of abnormal segregates in F_2 will be considered again in the next chapter.

We have now found evidence of interspecific differences corresponding to our first three types of inter-varietal difference. It will be remembered that the fourth type was a difference in the order in which genes are arranged in the chromosome. This has been demonstrated as between *Drosophila melanogaster* and the related species, *D. simulans*. They can be crossed, but give sterile hybrids. But the possibility of hybridisation renders it possible to homologise genes with more certainty than in rodents. Thus it is a matter of faith that a cross between an albino rabbit and a guinea-pig, if it were possible, would give albinos, a matter of fact that the cross between white-eyed *melanogaster* and *simulans* gives white-eyed offspring. On the other hand, several genes with fairly similar effects were shown by this test not to correspond. The chromosome maps resulting from this analysis are shown in Fig. 4. It will be seen that in the course of evolution a piece of the third chromosome has got reversed, as it occasionally does in geographical races and X-rayed forms of *melanogaster*. In *Drosophila obscura* there are five pairs of chromosomes, instead of the four pairs of *melanogaster*. Hybridisation is impossible, but a number of genes very clearly correspond in the two species. It turns out that the X chromosome of *obscura* is about twice as long as that of *melanogaster*. A group of four genes which lie at one end in *melanogaster* lie near the middle in *obscura*, in the same order, and the other half of its X chromosome seems to correspond to part of the third in *melanogaster*. Similarly in the rodents C and P are linked in the mouse and rat, but not in the guinea-pig. E is sex-linked in the cat (hence the difficulty of obtaining tortoiseshell tom-cats) but not in any other mammals so far investigated. In the course of evolution, then, there must have been a considerable amount of rearrangement of the material from which the chromosomes are built up, and which is the physical basis of heredity.

We have presumptive evidence that a rearrangement of the materials of the sex chromosomes is very common. In 1922 I (Haldane, 1922) formulated the following law: "When in the first generation between hybrids between two species, one sex is absent, rare, or sterile, that sex is always the heterogametic sex"—that is to say, the sex which produces two sorts of gametes, namely the male in most animal groups, the female in birds and *Lepidoptera*. This rule was formulated on the basis of forty-eight agreements and one exception. Since then Crew and others have found a number of further agreements and no further exceptions. When I formulated the law in question I attempted to explain it, but the explanation was somewhat inadequate. Since Stern has produced the condition experimentally within a species (p. 48), I regard his explanation as probably valid for most, if not all, of my cases, which are therefore due to interspecific differences in the sex chromosomes.

A concrete example was found by Lancefield (1929), who worked with *Drosophila obscura*. He found two morphologically indistinguishable races or subspecies of this species, whose habitats overlapped, although they were not quite identical, and there was a moderate psychological obstacle to crossing, perhaps a matter of odour. When they were crossed the males were sterile but the females fertile. Cytological examination showed that the Y chromosome was twice as large in one race as the other.

At this point it becomes necessary to consider the extraordinary, and so far unique, state of affairs found in the genus *Oenothera*. Some of its species, including *Oenothera Hookeri*, contain fourteen chromosomes which pair regularly, and the species breeds true. In others, such as *Oenothera Lamarckiana*, the conditions are very different. Only two of the chromosomes pair before reduction. The other twelve form a ring, in which alternate members go to each pole, like men and women who have been dancing alternately in a ring, and then separate. When *Lamarckiana* and *Hookeri* are crossed we get two sharply different kinds of hybrid, so that *Lamarckiana* turns out to be a permanently heterozygous organism. It does not, however, when self-fertilised give a progeny in the ratios 1:2:1 like an ordinary heterozygote. Both the homozygous types die before germination. Occasionally, however, crossing-over occurs, giving rise to a new type which is viable. About 2 per cent. of the seedlings of *Lamarckiana* selfed differ from it, the cause being sometimes rearrangement due to crossing-over, but more frequently the presence of an extra chromosome. De Vries, who discovered the phenomenon, called these abnormal plants mutants. It is clear that they are quite different in their origin and behaviour from the far rarer mutants of ordinary species, and throw little light on the general problem of evolution. The ring-forming species of *Oenothera* seem to have been evolved from the normal species by a series of interchanges between different chromosomes, such as we saw occurred in *Pisum*. The differences between different *Oenothera* species depend mainly on this interchange. They have most recently been analysed by Darlington (1931), but a condensed account is hardly possible.

It is extremely common among plants to find groups of closely related species where the numbers of chromosomes are simply related. Thus in the genus *Chrysanthemum*, the number of chromosomes going into a gamete in nineteen species is 9, 18, 27, 36 or 45, all multiples of 9. Similarly in *Rosa* we have a series of multiples of 7, in Prunus of 8, in Salix (with one exception) multiples of 19. Clearly the process of species-formation in these cases must have been sudden.

It is an important fact that many of our most valuable cultivated plants have two or three times the chromosome numbers of related species. These include wheat, oats (but not barley, maize, and rice), plums, many cherries, most roses, the dahlia, and many others. Now, most of these polyploid plants, when their genetics are investigated, do not behave like the autopolyploid *Primula sinensis*, where two chosen at random out of a set of four homologous genes go into a gamete. They resemble rather the allopolyploid *Primula kewensis.* And their cytology shows that each chromosome has a definite mate. They unite in pairs, not in fours or sixes such as are found in an autoploid, though a moderate amount of secondary association is not rare. There can, I think, be little doubt that the forms with the smaller chromosome number are the more primitive. The most obvious theory is that a species with, say, twenty-eight chromosomes, like the hard wheats (*Triticum durum* and related forms), has arisen by hybridisation of two fourteen-chromosome species and subsequent doubling. However, Percival, our greatest authority on wheats, in his classical book "The Wheat Plant" (1921), held that the twenty-eight-chromosome hard wheats are probably autopolyploids of the fourteen-chromosome wild small spelt, *Triticum monococcum*. It may be that in the course of ten thousand years or so (a negligible period from the point of view of evolution) an autopolyploid would be able to evolve so that each chromosome had one and only one definite mate. On the other hand, the third set of chromosomes found in the bread wheats such as *Triticum vulgare*, with forty-two chromosomes, almost certainly comes from the wild grass *Aegilops ovata*, or some nearly related species. This was conjectured by Percival on morphological grounds, but has since been made almost certain by the studies of Kihara (1929) and Percival (1930) on *Triticum-Aegilops* hybrids. Unfortunately time does not permit me to dwell on the entirely fresh light which Vaviloff (1926) has thrown on human prehistory by his studies of the wheat plant. But I cannot pass them over completely. When crossing is possible one can, as in the case of wheat, obtain a little evidence as to the origin of the sets of chromosomes in natural polyploids. Where it is impossible comparative morphology may give us strong indications.

Until quite recently we had no analogy among interspecific differences to the fifth type of intervarietal difference described in the last chapter—namely, the reduplication of some, but not all, of the chromosomes. This has recently been found by Darlington and Moffet (1930). In the *Rosaceae* the basic number is generally seven, as in the various species of *Rosa* and *Rubus*. In the apple, *Pyrus malus*, and some related wild species, the haploid number is seventeen, the somatic number is being thirty-four or fifty-one. But in the formation of pollen in a diploid the chromosomes do more than pair. They associate in groups which may include four groups of four and three groups of six similar chromosomes. Calling the gametic complement of *Rosa* or *Rubus* abcdefg, that of *Pyrus* seems to be :

> abcdefg,
> abcdefg,
> abc.

Thus three of the chromosomes are represented six times in the zygote, four of them four times. The ordinary balance is clearly upset. When this happens within a species, as in the trisomic mutants of *Datura*, the plant so produced is generally a little weaker than the original type, and moreover does not breed true. We must suppose that in the evolution of *Pyrus* from some form like *Rosa* or *Potentilla*, which has basal haploid numbers of seven and fourteen respectively, part of a set of chromosomes was either dropped or reduplicated. The new balance proved viable, and gradually the chromosomes found definite mates so that pairing became regular.

A few words may be said on other types of species cross which throw little light on the nature of specific differences. When *Nicotiana tabacum* and *N. sylvestris* are crossed (Goodspeed and Clausen, 1922), the hybrid is very like the variety of *N. tabacum* employed. This is natural enough, as *tabacum* has forty-eight chromosomes and *sylvestris* only twenty-four. The hybrids, being triploids, are very sterile, but if carefully pollinated may give about 1 per cent. of the normal number of seeds. Crossed with *tabacum* pollen we get a fair variety of forms, but all pretty like *tabacum*, and from these we can get races indistinguishable from *N. tabacum*. Crossed with *sylvestris* most of the seedlings are monstrous, but about 10 per cent. resemble *sylvestris* and breed true. Clearly the only functional gametes are those which contain a set of genes almost the same as those of the parent species. This type of hybrid behaviour is unfortunately useless for analysing the nature of the interspecific differences. But in nature it may serve occasionally to bring across a gene from one species to another.

To sum up, interspecific differences are of the same nature as intervarietal. But the latter are generally due to a few genes with relatively large effects, and rarely to differences involving whole chromosomes or large parts of them. The reverse is true of differences between species. The number of genes involved is often great, and cytologically observable differences common. It is largely these latter which are the causes of interspecific sterility.

CHAPTER IV

Natural Selection

"Per uarios casus, per tot discrimina leti
Tendimus ad Latium."

Vergil, *Aeneid.*

Before we discuss natural selection, it will be well to consider populations which are in equilibrium, although several different genotypes exist. I will give two examples out of many. The common snails, *Cepea hortensis* and *C. nemoralis* (the *Helix* of our childhoods) possess several varieties which differ as regards the banding on their shells. They are due to the action of several genes which are multiple allelomorphs, or at least very strongly linked. In the various breeding experiments done on them by Lang (1911) and others, Mendel's laws were qualitatively obeyed, *i.e.* no mutations appeared. So mutation is infrequent. Diver, in very careful unpublished work which he kindly allows me to quote, found no selective destruction of any type by birds. A typical population to-day consists of rather more banded than unbanded individuals.

Now deposits of these snail shells exist going back to the Red Crag Age in England and the Miocene in France. In early Iron Age and Neolithic deposits, the types are found in about the present proportions. In Pleistocene deposits the two types are more nearly equal in number (Diver, 1929). So the population, though polymorphic, is very stable. One cannot be sure that it has altered significantly in so short a time as a quarter of a million years.

In man a series of three multiple allelomorphs divides us into four blood groups, which determine whether or not blood can safely be transfused from one individual into another. The proportions of these genes are characteristic of different races.

The Jews of Salonica had lived there as an endogamous community for over four centuries after their expulsion from Spain, but still resembled Arabs rather than their Greek neighbours in 1918. Similarly the Germans in Hungary resembled their relations in Germany, the gypsies being like Indians. Clearly selection affects the proportions of these characters very slowly, and mutation is known to be much too rare to produce appreciable effects in a thousand years. Any given people is therefore nearly in equilibrium.

If mating in any population is at random, it reaches equilibrium within a single generation as regards the proportions carrying two, one, or none of a given autosomal (*i.e.* non-sex-linked) gene. The ratios are:

$$u\mathrm{AA} : 2\, u\mathrm{Aa} : 1\ aa.$$

For example, if one man in ten thousand is an albino, $u = 99$, and nearly one person in fifty is heterozygous for albinism. Actually these conditions are not quite fulfilled. Marriages of cousins are much commoner than would be the case if we married at random, and about a sixth of human albinos are the offspring of such unions. So albinism is commoner than it would be were mating strictly at random. But it is important to realise that the proportion of the population bearing a recessive character shows no tendency to diminish further after the first generation of random mating, unless the character is disadvantageous. Extra-nuclear factors behave in a similar way. Apart from selection or mutation, they do not tend to spread. The same applies to chromosomal abnormalities unless they interfere with the normal reduction mechanism, or lower the fertility or viability of the gametes or zygotes carrying them, which, however, they very often do.

Before, however, we deal with the theoretical effects of natural selection, it will be well to give a few examples of it, because the statement is still occasionally made that no one has actually observed it at work. It is quite true that the observations so far made are far from adequate. But at least they prove the existence of natural selection as a fact. In a random mating species matters are complicated because type A produces type B offspring, and so on.

TABLE III *Percentages belonging to the Four Blood Groups*

	AB	A	B	O
Germans in Heidelberg	5	43	12	40
Germans ⎤	3	43	13	41
Magyars ⎬ in Hungary	12	39	19	31
Gypsies ⎦	6	21	39	34
Indians in Northern India	9	19	41	31

Hence observations have often been confined to one part of the life cycle. For example, di Cesnola (1904), tied up twenty green and forty-five brown specimens of the insect *Mantis religiosa* with silk threads in green grass, and found that thirty-five of the brown and none of the green had been eaten by birds in nineteen days. Twenty browns and twenty-five green were tethered in brown grass, all the greens were dead in eleven days, five having been killed by ants. The browns were all alive at the end of nineteen days. Clearly protective colouring is a reality in this case.

More satisfactory results are obtained in organisms which are either self-fertilising or apogamous, so that type A produces only type A offspring, and so on. Here the best work has been done in Russia, where Darwinism, being part of the official creed, is a much more vital question than in other countries. A good deal has been done on cereals, in answer to the question of what becomes of the progeny of a given mixed batch of seed when it is harvested in several successive years. Fig. 7, from Sapegin's work, shows a typical example. It will be seen that, under these artificial conditions, selection is extremely intense. The whole question of natural selection under cultural conditions has been taken up by the Russian school. It appears that under these conditions there is mimicry like the well-known mimicry of natural species. Thus Russian flax seed commonly contains seeds of *Camelina linicola*, which mimics flax both in habit and in size and shape of its seeds. The seeds of *Camelina glabrata*, which is probably an ancestral form, are smaller, and presumably unconscious and unwilling human selection has picked out the largest seeded types and thus modified the species. Berg (pp. 324–328) gives seven similar cases.

It may be argued that we are here dealing with artificial, rather than natural selection. I think such a criticism can only apply when selection is deliberate. Apart from such cases, man is merely creating a new type of environment. Thus when the present breakwater was built across the mouth of Plymouth harbour the water inside became on the whole muddier, and the crab *Carcinus maenas* developed roomier gill chambers (Weldon, 1895). This may not have been due to natural selection, but it was certainly not a case of artificial selection.

In many ways the work of Sukatschew (1928) on dandelion (*Taraxacum officinale*) is more striking. This species consists of a number of pure lines which breed perfectly true, and do not cross, as they propagate by apogamy. Sukatschew worked with six lines:

A, B, and C from the same lawn at Leningrad, X from Archangelsk in the far north, Y from Vologda, a little south but a long way east of Leningrad, and Z from the Crimea.

The plants were morphologically distinguishable. Thus A had the most finely divided leaf, B the hairiest, and C was the tallest and had red petioles. Large numbers of the six types were grown from seed, and then planted out. They were planted in two densities, either 18 or 3 centimetres apart. The latter arrangement gives thirty-six times as many plants on a given area as the former.

Z, the Crimean type, was a hopeless failure in Leningrad. Even on the sparsely sown plots over 50 per cent. died in the first year, as compared with none of A, C and Y, 4 per cent. of X, and 10 per

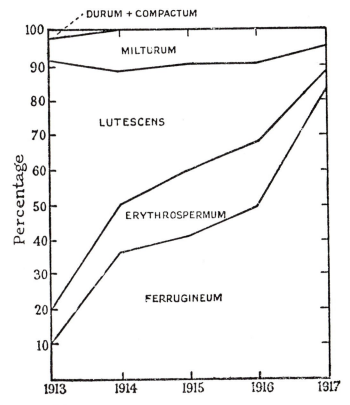

Figure 7. Percentages of wheats of different types in a mixed population in successive years (after Sapegin, 1922). No artificial selection was practised.

cent. of B. The others competed more equally. The results of competition between A, B and C are given in Table IV.*

As it was necessary to pull up the plants in the mixed cultures to distinguish the races, no counts were made in the first year. Duplicate experiments agreed. It will be seen that in three cases out of four the order of viability was C, A, B, but in the pure dense cultures the order was reversed. In order, however, to estimate the fitness, in a Darwinian sense, of the three races, we must go further, and compare the fertility. In the sparse cultures A produced on an average twenty-seven flowers per plant, B thirty-eight, and C ten. A and B produced about seventy seeds per flower, C about a hundred and forty. Thus B was probably the most successful type, though it had the greatest mortality. But in the dense mixed cultures C was not only the most viable, but the best cropper.

Thus the "fitness" depends in a quite complicated way on the environment. In order to test fitness in the Darwinian sense it would have been necessary to grow the plants in competition and in presence of grass in a plot covered in with a gauze roof to prevent the entrance of foreign seed. Quite possibly in the presence of grass the order of fitness might have been different.

Similar cases are recorded by Engledow (1925) in wheat and Sax (1926) in beans. Engledow found that when two wheats, Red Fife and Hybrid H, were spaced at 2 inches by 2 inches the former yields the larger crop; at 2 inches by 6 inches the yields are nearly equal. At greater distances Hybrid H is the better cropper. Sax compared bean races differing in respect of a gene for colour. The whites

TABLE IV[1]

Race		Per cent Dead			
		First Year		Second Year	
		Sparse	Dense	Sparse	Dense
Pure Cultures	A	0	70	22.9	73.2
	B	10.3	19	31.1	51.1
	C	0	3.5	10.3	75.9
Mixed Cultures	A	–	–	16.5	77.4
	B	–	–	22.1	80.4
	C	–	–	5.5	42.0

[1] From Sukatschew.

* Correction to the original text from Table II to IV.

generally gave a smaller crop, but in very favourable conditions a larger.

The best example known to me where the effects of selection have been watched over many generations is described by Todd (1930), in the case of the organism of which different races cause scarlet fever, puerperal fever, and erysipelas, *Streptococcus haemolyticus*. It rapidly loses its virulence for animals when bred in artificial media. This phenomenon was at first taken for a Lamarckian inheritance of the effects of disuse, and its analysis by Todd is typical of the results obtained when such phenomena are carefully studied. He found that his *Streptococcus* when grown on agar produced hydrogen peroxide. But occasionally a variant appears which gives rather different colonies, is less virulent, and produces much less peroxide. We do not know how these variants arise because the details of the process of reproduction in bacteria are not known. There is no reason to think that bacterial mutation is a phenomenon essentially different from mutation in higher organisms, and it is not even clear that it is commoner.

Now the normal type of bacteria, when grown on agar, make enough peroxide to kill themselves, or at any rate to slow down their growth very greatly. When parasitic they are protected by the catalase of their hosts. This is a widely distributed enzyme which destroys hydrogen peroxide. Hence the glossy and non-virulent type is the only survivor after a few weeks of culture. But if a little catalase is added to the medium the virulent type grow as well as the non-virulent, and can be preserved in culture indefinitely. These bacteria divide about once every half-hour, so Todd's experiments, which lasted thirty-nine days at a minimum, covered some 2000 bacterial generations, corresponding to about 50,000 years in human evolution, and a century even with so rapidly breeding a creature as *Drosophila*. It took Calmette and Guerin (1924) fourteen years, or about 25,000 generations, to convert the bovine tubercle bacillus into a harmless and indeed beneficial organism by growing it on artificial media. There is thus no reason to put down such modifications of bacteria to anything but natural selection, acting on the results of mutation.

The following example, from the work of Harrison (1920), shows natural selection at work among the moths of the species *Oporabia autumnata*. About 1800 a mixed wood of pine, birch, and alder on Eston Moor in Yorkshire was divided into two parts separated by half a mile of heather. In 1885, after a storm, the pines were replaced by birch in the southern portion, while in the northern birches and alders are now rare. Presumably in 1800 the two populations were similar. By 1907 they were quite different. In the pine-wood 96 per cent. belong to a dark variety, 4 per cent. to a light. In the birch-wood about 15 per cent. are dark and 85 per cent. light. The reason for this is fairly clear. In the pine-wood owls, nightjars, and bats feed on the moths, leaving their wings when the bodies are eaten. Although only 4 per cent. of the moths in the pine-wood are light, the majority of the wings lying on its grass belong to the light variety, which is thus some thirty times as likely to be caught as the dark. We do not know for certain what advantage the light-coloured insect enjoys in the birch-wood, where birds and bats are relatively rare. But as the light race lays its eggs later than the dark they are less likely to hatch in the same year instead of (as normally) in the spring, an event which entails the death of the larvae during the winter. It may be added that an attempt to make the pine-wood insects lighter in colour by feeding them for three generations on birch met with no success.

It is perfectly true, as critics of Darwinism never tire of pointing out, that in these observations no new character appears in the species as the result of selection. Novelty is only brought about by selection as the result of the combination of previously rare characters. Supposing that in a population fifteen characters, not correlated, are each present in 1 per cent. of the individuals. The combination of all fifteen would only be present in one in 10^{30}, *i.e.* 1 per (English, not American) quintillion. This is a large number even on the evolutionary scale. The earth's land surface is only 10^{18} square centimetres. There have not been 10^{30} higher plants in the whole of geological history (10^9 years), including all members of all many-celled plant species. The combination of all fifteen characters would not occur in practice even once in the whole history of plants larger than unicellular.

Now suppose that natural selection acts on all these fifteen characters, so that they are found in 99 per cent., not 1 per cent. of the species. The combination of all fifteen would now be found in 86 per cent. of the population. It would, in fact, be the normal character. No one has ever observed this happening in nature, because, owing to the slowness of natural selection, it would probably require ten thousand years of observation in a favourable case. But, as Darwin realised, it has happened as the result of artificial selection. A middle-white pig differs from a wild boar in some thirty to forty distinct respects. Some may be due to the action of the same gene on several organs. Others require several genes. Some of these genes were doubtless present in rarities in the wild species. Others may have turned up after domestication, but if so they had probably often occurred in the wild species.

It is important to realise that the combination of several genes may give a result quite unlike the mere summation of their effects one at a time. This is obviously to be expected if genes act chemically. Thus in *Primula sinensis* a dark stem (recessive) is associated with no great change in colour of acid-sapped (red and purple) flowers. But blue (recessive) flowers, which have a neutral sap, when growing on a dark stem, are mottled. The same recessive dark stem genes, along with genes for a green stem, give plants which will not set seed, though they give good pollen. So selection acting on several characters leads, not merely to novelty, but to novelty of a kind unpredictable with our present scientific knowledge, though probably susceptible of a fairly straightforward biochemical explanation.

We have seen that there is no question that natural selection does occur. We must next consider what would be the effect of selection of a given intensity. The mathematical theory of natural selection where inheritance is Mendelian has been mainly developed by R. A. Fisher, S. Wright, and myself. Some of the more important results are summarised in the Appendix, but I shall deal with a few of them here. The first question which arises is how we are to measure that intensity. I shall confine myself to organsims, such as annual plants and insects, where generations do not overlap. The more general case, exemplified by man, can only be treated by means of integral equations. Suppose we have two competing types A and B, say dark and light moths or virulent and non-virulent bacteria. Then if in one generation the ratio of A to B changes from r to $r(1 + k)$ we shall call k the coefficient of selection. Of course k will not be steady. In one year an early spring will give an advantage to early maturing seeds. In the next year a late frost will reverse the process. Nor will it be constant from one locality to another, as is clear in the case of the moths just cited. We must take average values over considerable periods and areas. The value of k will increase with the proportion of individuals killed off by selection, but after selection has become intense enough to kill off about 80 per cent. of the population it increases rather slowly, roughly as the logarithm of the number killed off per survivor—sometimes even as the square root of the logarithm. In what follows I shall suppose k to be small.

The effect of selection of a given intensity depends entirely on the type of inheritance of the character selected and the system of mating. I will confine myself for the moment to characters inherited in an alternative manner, in a population either mating at random or self-fertilised. If two races do not cross, or if the inheritance is cytoplasmic, and if u_n is the ratio of A to B after n generations, then $u_n = e^{kn}u_o$, or $kn = \log_e (u_n/u_o)$. If the character is due to a single dominant gene, and u_n is the ratio of dominant to recessive genes, then

$$k_n = u_n - u_o + \log \frac{u_n}{u_o}.$$

This means that selection is rapid when populations contain a reasonable proportion of recessives, but excessively slow, in either direction, when recessives are very rare (see Fig. 8). Thus if $k = \frac{1}{1000}$ i.e. 1001 of one type survive to breed for every 1000 of the other, it would take 11,739 generations to increase the number of dominants from one in a million to one in two, but 321,444 generations to increase the number of recessives in the same way. It is not surprising that the only new types which have been known to spread through a wild population under constant observation are dominants. For example, the black form of the peppered moth, *Amphidasys betularia*, which replaced the original form in the industrial districts of England and Germany during the nineteenth century, is a dominant. When the character is due to several rare genes the effect of selection is also very slow even if the genes are dominant. But however small may be the selective advantage the new character will spread, provided it is present in enough individuals of a population to prevent its disappearance by mere random extinction. Fisher has shown that it is only when k is less than the reciprocal of the number of the whole population that natural selection ceases to be effective. An average advantage of one in a million will be quite effective in most species.

A curious situation arises when two genes one at a time produce a disadvantageous type, but taken together are useful. Such a case was found by Gonzalez (1923), with three of the well-known genes in *Drosophila*. It will be seen from Table V that two of the genes, Purple and Arc, lower the expectation of life in both sexes. The third, Speck, increases the expectation of life in males, without altering it significantly in females. Purple and Arc together give considerably longer life in both sexes, but especially in the male, than either alone. The combination of all three genes restores the normal duration of life in both sexes, the increase being insignificant.

The figures for progeny in the last column are based on few families, but the fertility of Purple is significantly greater than that of the wild type. If the percentage of fertile matings is not greatly lowered by this gene, it would tend to spread in a mixed population

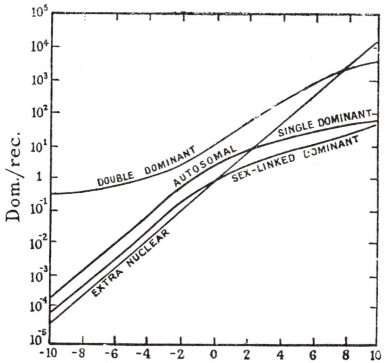

Figure 8. Theoretical results of selection on the composition of a population when dominants are favoured. Abscissa, number of generations multiplied by coefficient of selection. Ordinate, ratio of dominants to recessives. If the races do not interbreed the effect is the same as for an extra-nuclear factor. In the case of the double dominant, the genes are supposed to be present in equal numbers. If recessives are favoured, the sign of the abscissa is changed. For example, if $k = 0.01$, it will be seen that about 400 generations are needed for the ratio of dominants to recessives to change from 1 to 10, if autosomal single dominants are favoured.

under Gonzalez' cultural conditions, though doubtless in a state of nature this is not so.

Of course the life-lengths of Table V do not represent selective advantages, but they only refer to five of the hundreds of genes known in *Drosophila*. No doubt none of the other common mutant genes *by itself* is advantageous in nature, or it would have spread through the species and established itself. But it is quite possible that a combination of two, three, or more would be so. The number of possible combinations of all the known genes is very large indeed. The combined mass of a population consisting of one fly of each possible type would vastly exceed that of all the known heavenly bodies, or that of the universe on the theory of general relativity. It is not an extravagant theory that at least one member of this population would be better adapted for life than the present wild type.

If we consider a case where the double dominant AB and the double recessive *aabb* are both more viable than the types A*bb* or *aa*B, then a population consisting mainly of either of the favoured types is in equilibrium (see pp. 107–108), and mutation on a moderate scale is not capable of upsetting this equilibrium. But the change from one stable equilibrium to the other may take place as the result of the isolation of a small unrepresentative group of the population, a temporary change in the environment which alters the relative viability of different types, or in several other ways, one of which will be considered later.

This case seems to me very important, because it is probably the basis of progressive evolution of many organs and functions in higher animals, and of the break-up of one species into several. For

TABLE V *Mean Life in Days, and Average Progeny per Fertile Mating of several types of Drosophila melanogaster*

Type	Life of ♂	Life of ♀	Progeny
Wild	38.08 ± 0.36	40.62 ± 0.42	247
Purple (eyes)	27.42 ± 0.27	21.83 ± 0.23	325
Arc (wings)	25.20 ± 0.33	28.24 ± 0.37	127
Speck (in axilla)	46.63 ± 0.63	38.91 ± 0.65	103
Purple arc	36.00 ± 0.53	31.98 ± 0.43	230
Purple speck	23.72 ± 0.22	22.96 ± 0.19	247
Arc speck	38.41 ± 0.58	34.69 ± 0.66	106
Purple arc speck	38.38 ± 0.62	40.67 ± 0.45	118

an evolutionary progress to take place in a highly specialised organ such as the human eye or hand a number of changes must take place simultaneously. Thus if the eye is unusually long from back to front we get shortsightedness, which would not, however, occur if there were a simultaneous decrease in the curvature of the cornea or lens, which would correct the focus. As, however, abnormal eye-length is fairly common, being often inherited as a dominant, while lessened corneal curvature is rare, the usual result of the condition is short-sightedness. Actually a serious improvement in the eye would involve a simultaneous change in many of its specifications. Occasionally a single gene might produce simultaneous and harmonious changes in many at once, but this is not generally the case with new mutants, although some such genes, being almost harmless, are not eliminated, and account for much of the variation in natural populations. Evolution must have involved the simultaneous change in many genes, which doubtless accounts for its slowness. Here matters would have been easier if heritable variations really formed a continuum, as Darwin apparently thought, *i.e.* if there were no limit to the possible smallness of a variation. But this is clearly not the case when we are considering meristic characters. Mammals have a definite number of neck vertebrae and chromosomes, most flowers a definite number of petals, exceptional organisms being unhealthy. And the atomic nature of Mendelian inheritance suggests very strongly that even where variation is apparently continuous this appearance is deceptive. On any chemical theory of the nature of genes this must be so.

If the only available genes produce rather large changes, disadvantageous one at a time, then it seems to me probable that evolution will not occur in a random mating population. In a self-fertilised or highly inbred species it may do so if several mutations useful in conjunction, but separately harmful, occur simultaneously. Such an event is rare, but must happen reasonably often in wheat, of which the world's population is roughly 5×10^{14} plants, about 99 per cent. of which arise from self-fertilisation. But where natural selection slackens, new forms may arise which would not survive under more rigid competition, and many ultimately hardy combinations will thus have a chance of arising. Ford (1930) describes a case which may be interpreted in this way in the butterfly *Melitaea aurinia*. This seems to have happened on several occasions when a successful evolutionary step rendered a new type of organism possible, and the pressure of natural selection was temporarily slackened. Thus the distinction between the principal mammalian orders seems to have arisen during an orgy of variation in the early Eocene which followed the doom of the great reptiles, and the establishment of the mammals as the dominant terrestrial group. Since that date mammalian evolution has been a slower affair, largely a progressive improvement of the types originally laid down in the Eocene.

Another possible mode of making rapid evolutionary jumps is by hybridisation. As we saw, this may lead to the immediate formation of a new species by allopolyploidy. An example of this process in Nature is given by Huskins (1930). The rice-grass *Spartina Townsendii* first appeared on the muddy foreshore of Southampton Water about 1870. It breeds true, but Stapf (1927) regards it as a hybrid of the English *S. stricta* and the (probably) American *S. alterniflora*. Huskins finds that these two latter species have chromosome numbers of 56 and 70 respectively, while *S. Townsendii* has 126 chromosomes. The basic chromosome number in the Gramineae is 7, so it would seem that a cross between an octaploid and a decaploid species gave rise to an enneaploid with 63 chromosomes, vigorous, but somewhat sterile and not breeding true. The doubling of its chromosome number gave an octocaidecaploid which combined hybrid vigour with fertility and stability. This interpretation must of course remain doubtful until the crossing has been repeated under controlled conditions, but the conjunction of morphological and cytological evidence renders it very likely.

Meanwhile the new species is proving its fitness in a true Darwinian manner by exterminating its parents, and also according to the ideas of Kropotkin, by aiding the Dutch in their struggle with the sea. Its recent origin is to be explained by the fact that its parents only hybridised as the result of human activity, *S. alterniflora* having presumably been brought on a ship from America.

Apart from this, hybridisation (where the hybrids are fertile) usually causes an epidemic of variation in the second generation which may include new and valuable types which could not have arisen within a species by slower evolution. The reason for this is that genes often exhibit quite novel behaviour in a new environment. Thus Kosswig (1929) crossed the fishes *Platypoecilus* and *Xiphophorus*, and found that some, though not all, of the genes causing abnormal colours in the latter produced very exaggerated effects when introduced into the former. Thus a gene from *Xiphophorus* for black pigment produced hybrid fish which, though quite healthy, were covered with warts of black pigment. Lotsy (1916), in particular, has emphasised the importance of hybridisation in evolution, and shown that it occurs in nature. He was able, for example, by crossing two species of *Antirrhinum* (snapdragon), namely *A. majus* and *A. glutinosum*, to obtain in the second generation plants whose flower would be ascribed by a taxonomist to the related genus *Rhinanthus*. At one time Lotsy did not believe in mutation, except by loss, and attributed all variation to hybridisation. This is certainly an exaggeration. Not only has mutation now been fully confirmed, but no such hypothesis as Lotsy's will explain the slow and steady evolution to which the geological record bears witness. Nevertheless it is difficult to doubt that hybridisation has rendered possible the coming together of certain combinations of genes which could not have arisen otherwise.

Still another possible way out of the impasse is as follows. Instead of the two or more genes changing abruptly, they may change in a number of small steps, *i.e.* multiple allelomorphs may appear causing very slight changes in the original type of gene. Supposing blackness conferred a small advantage of about one in a thousand on the wild mouse by acting as a protective colour or otherwise, it would not be favoured by selection because it confers a definite physiological handicap. The death-rate among black mice in their first three weeks of life was shown by Detlefsen and Roberts (1918), to be decidedly larger than that of the wild type (two other colour genes had no such effect). Actually the handicap was about $4\frac{1}{2}$ per cent. But it might pay the mouse to become slightly darker, changing its G gene a fraction of the way towards the gene producing black, to wait until modifying genes had accumulated which restored the physiological balance, then to proceed another step, and so on.

It will be remembered that Detlefsen (1914) crossed the ordinary guinea-pig, *Cavia porcellus*, with the smaller and darker coloured *Cavia rufescens*. He found that the dark colour was mainly due to a modification of the gene G, which behaved as a multiple allelomorph with the gene for black. Thus if *Cavia porcellus* represents the original type, which is highly probable, its gene G has changed part of the way towards producing blackness in the evolution of *C. rufescens*. Three other crosses between subspecies and geographical races in rodents give similar results. It looks as if the evolution of colour in rodents generally proceeded by rather small steps. My own quite speculative theory of orthogenetic evolution such as that described in Chapter I is that we are dealing here not only with the accumulation of numbers of genes having a similar action, but with the very slow modification of single genes, each changing in turn into a series of multiple allelomorphs. The phrase "modification of the gene" is of course a rather misleading simplification. What I mean is that mutation was constantly modifying the gene, and that at any given time natural selection acted so as to favour one particular grade of modification at the expense of the others.

One more application of mathematics, and I have done. Under what conditions can mutation overcome selection? This is quite a simple problem. Let p be the probability that a gene will mutate in a generation. We saw that p is probably usually less than a millionth, and so far always less than a thousandth. Let k be the coefficient of selection measuring the selective disadvantage of the new type, k being considerably larger than p. Then equilibrium is reached when the proportion of unfavourable to favourable phenotypes is p/k if the mutant is recessive, $2p/k$ if it is dominant. The above calculations refer to a random mating population. The ratio is always p/k in a self-fertilised population. Hence, unless k is so small as to be of the same order as p, the new type will not spread to any significant extent. Even under the extreme conditions of Muller's X-ray experiments, when mutation was a hundred and fifty times more frequent than in the normal, a disadvantage of one in two thousand would have kept any of the new recessive types quite rare. Thus until it has been shown that anywhere in nature conditions produce a mutation rate considerably higher than this, we cannot regard mutation as a cause likely by itself to cause large changes in a species. But I am not suggesting for a moment that selection alone can have any effect at all. The material on which selection acts must be supplied by mutation.

Neither of these processes alone can furnish a basis for prolonged evolution. Selection alone may produce considerable changes in a highly mixed population. A selector of sufficient knowledge and power might perhaps obtain from the genes at present available in the human species a race combining an average intellect equal to that of Shakespeare with the stature of Carnera. But he could not produce a race of angels. For the moral character or for the wings he would have to await or produce suitable mutations.

CHAPTER V

What is Fitness?

"I returned, and saw under the sun that the race is not to the swift, nor the battle to the strong, neither yet bread to the wise, nor yet riches to men of understanding, nor yet favour to men of skill; but time and chance happeneth to them all."—*Ecclesiastes.*

We have seen that natural selection is a reality, that the facts of variation, though different from what Darwin believed them to be, are yet such as to yield a raw material on which natural selection can work. We have also seen that variation directly induced by the environment is not in itself competent to explain the known facts of evolution. But we know very little about what is actually selected, and any attempt to give a concrete account of natural selection at work must be decidedly speculative. Nevertheless such an attempt must be made. I believe that the opposition to Darwinism is largely due to a failure to appreciate the extraordinary subtlety of the principle of natural selection. Attention has so far been focused (and inevitably so) on the crudest type of interaction between organisms and environment.

Lucretius stated the principle of natural selection in its crudest form when he wrote:

"Multaque tum interiisse animantum saecla necesse est,
Nec potuisse propagando producere prolem.
Nam quaecomque uides uesci uitalibus aureis,
Aut dolus, aut uirtus, aut denique mobilitas est
Ex ineunte aeuo genus id tutata, reseruans."[1]

This was a magnificent intellectual discovery when Lucretius made it two thousand years ago. But it is only a small part of the whole story.

---

[1] "And many lines of organisms must have perished then, and been unable to propagate their kind. For whatever you see feeding on the vital air, either craft, strength, or finally mobility has been protecting and preserving that race from its earliest times."

We must look a little deeper. There is a perfect analogy in the field of history. Primitive history is largely an account of battles. The state is considered in its most obvious relationship to other states, and nothing is said about its internal structure apart from the crudest outlines.

"Thine, Roman, is the pilum, Roman, the sword is thine,
The even trench, the bristling mound, the legion's ordered line,"

is the explanation of the greatness of Rome served out to us in our infancy. We soon realise its inadequacy. We find that we can obtain a better idea of the real greatness of the Roman character from the ideal Roman, Aeneas, or the almost equally mythical heroes of Livy. Still later we realise the vast complexity of the problem, the extraordinary concatenation of racial, economic, and traditional influences which must have co-operated to build Rome. "Tantae molis erat Romanum condere gentem."

So with an animal or plant. We are first struck by its obvious adaptations; its claws, teeth, spines, protective colouring, and so on. Such features impress the morphologist who is bound to note them when engaged in taxonomic work. But there remain a host of morphological characters which have no obvious value to their possessor. Such are the innumerable slight variations of leaf shape which often distinguish species of the same plant genus.

Later on we find subtler adaptations, for example, the high water-imbibing power of the colloids and the consequently low freezing-point of the cell-contents of cold-resisting plants, or the relation shown by Needham (1929) between the nitrogenous end-products of animal metabolism and the capacity of the embryo for getting rid of them.

But when we have pushed our analysis as far as possible, there is no doubt that innumerable characters show no sign of possessing selective value, and, moreover, these are exactly the characters which enable a taxonomist to distinguish one species from another. This has led many able zoologists and botanists to give up Darwinism. But before we follow their example it is desirable to consider certain facts.

Darwin himself was well aware of the correlation between different characters. To-day we see the same phenomenon as the multiple effects of a single gene. Since the gene exists in every cell of the body, it may be expected to affect the organism as a whole, even if its most striking effect is on some particular organ or function. Thus the gene C*h* in *Primula sinensis* incises the petals, doubles the number of sepals, breaks up the bracts, produces a more compact habit, increases the degree of crimping of the leaves when certain other genes are present, and so on. Morgan (1926) describes a number of cases of multiple action of this kind in *Drosophila*. In particular many genes modify wings and bristles simultaneously. Our actual description of a gene depends on our senses in a rather arbitrary way. Thus the gene for large central "eye" in the flower of *Primula sinensis* also shortens the style in plants where it would otherwise be long, producing a homostyle instead of a "pin" flower. If we were blind to the difference between yellow and white we should call it a modifier of the style length. In the same way in order to be hairy a stock (*Matthiola incana*) must have coloured flowers, and also carry two special genes for hairiness (Saunders, 1920). A race of blind botanists could have detected all four of the genes concerned, but would probably not know that two of them were concerned in producing certain aromatic compounds in the petals, which, though inodorous, incomprehensibly influenced insects in their visits. We are not much better off than these imaginary botanists. Most of the chemical constituents of living matter do not absorb visible radiation, we we cannot observe changes in them without painstaking analysis. To my mind it is probable that every gene produces a definite chemical effect, but we are very far from being able to prove this as yet.

Certainly we have no reason to offer why white-flowered beans should react more intensely to their environment, both in producing large and small crops, than those with coloured flowers; why white rabbits are not killed by an injection of nucleo-protein that will clot the blood in the vessels of normal rabbits, and so on. But these appear to be facts.

It is clear that most of the differences between species so far noted are very superficial. But along with them are differences of a kind which are much more important from the point of view of natural selection. Thus Crew (1927) describes a case of staphylococcal infection in mice which killed a whole group of Japanese waltzing mice, but no white European mice nor hybrids of the first generation. It also killed one-quarter of the F_2 generation and about half of the back-cross of the Japanese. One out of fifty-one in the offspring of the hybrids and European mice died. With this one exception, everything agrees with the view that immunity is due to a dominant gene found in the European mouse, absent in the Asiatic *Mus Wagneri*. Just the same straightforward Mendelian inheritance is found in the case of rust resistance in wheat. But presumably such genes will have effects even in the absence of disease. The fact that *Mus musculus* possesses the gene, doubtless owing to the extinction of those mice which lacked it, almost certainly has some effect on its other characters.

For such reasons I am not unduly impressed by the fact that the taxonomist's characters are not the useful aspects of the activity of those genes which distinguish species. Some of these genes may be needed for their obvious effects, but others will merely be required to restore a physiological balance. Thus Table V showed that if natural selection placed a premium, other things being equal, on arc wings in *Drosophila*, normal viability could be restored by the simultaneous appearance of purple eyes and an axillary spot. Perhaps some other genes would be equally effective, but it would require a combination of physiological and genetical research to determine the reason for their utility.

But in addition to characters of the type discussed above, we must be prepared to find within a species apparently useless characters which nevertheless have been selected for their own sakes, for an entirely different reason.

Bidder (1930) has stressed the importance in evolution of disasters which may occur only once in very many generations, but may have a profound selective effect. Thus three sponges which live normally on rocks between tide-marks have elaborate systems for ejecting water for great distances. These cannot be of any use to dwellers in surf. But Bidder points out that perhaps once in a century a violent rainstorm or great heat at a low spring tide will kill off all the population over a wide area, which will be replenished from those few members of the species which are living in caves or other sheltered spots where their canal-system is essential owing to the stagnation of the water. Similarly the apparently insane tendency to migration of certain rodent species may be due to the fact that they are all descended from unstably minded survivors of some great catastrophes, such as an ice-age in the past. Bidder's argument is generalised on p. 136.

One may apply similar arguments to plants. Plant dispersal is normally a very slow process. Thus Ridley (1905) studied the dispersal of the tropical tree *Shorea*. He found that the rather large but winged fruit might fly a hundred yards, and the tree so produced would take thirty years to grow to maturity. Thus about five hundred years would be needed for a migration of about a mile. This is the normal rate, but one seed in a thousand million carried a hundred miles by a typhoon or by accidental adherence to an animal would entirely upset such a calculation. For such reasons I believe that the task which the ecologists are now rightly setting themselves, of examining natural selection at work, may be harder than they think. Doubtless it will be possible to determine the normal incidence of selection in many species. But in a great disaster or a great migration the characters of the single survivor are what matters. And although such ecologists as Elton (1927) now realise that many animal species are periodically almost wiped out by disease, and may be able to determine the selective effect of these epidemics, they will hardly be able to study the single plant seed which once in a century, or more rarely, establishes a new species in a new continent. If it is a member of a polymorphic species it will establish a race differing in many respects from the normal, and here at least we must admit that mere chance is likely to play a certain part in species formation, though I think that Elton has somewhat exaggerated its importance.

But I must leave this fascinating topic to discuss a fallacy which is, I think, latent in most Darwinian arguments, and which has been responsible for a good deal of the poisonous nonsense which has been written on ethics in Darwin's name, especially in Germany before the war and in America and England since. The fallacy is that natural selection will always make an organism fitter in its struggle with the environment. This is clearly true when we consider the members of a rare and scattered species. It is only engaged in competing with other species, and in defending itself against inorganic nature. But as soon as a species becomes fairly dense matters are entirely different. Its members inevitably begin to compete with one another. I am not thinking only of the active and often conscious competition between higher animals, but also of the struggle for mere space which goes on between neighbouring plants of closely packed associations. And the results may be biologically advantageous for the individual, but ultimately disastrous for the species. The geological record is full of cases where the development of enormous horns and spines (sometimes in the male sex only) has been the prelude to extinction. It seems probable that in some of these cases the species literally sank under the weight of its own armaments. Again, while modern research tends to show that sexual selection in birds is rather less important in making bright colour and structures such as the peacock's tail advantageous in male birds than Darwin supposed, there is still a good deal of evidence that it has certain selective value in securing mates. And none will contend that (except in so far as it has induced Hindus to regard him as sacred and Europeans as a suitable pet) the peacock's rather cumbrous tail has been of any advantage to him in the struggle with the environment. I could multiply such cases indefinitely, but they are all somewhat uncertain. To prove our case we should want statistics as to the number of offspring left on the one hand, and expectation of life on the other, of long- and short-tailed peacocks under natural conditions.

I therefore propose to deal with a case where some at least of the required data do exist. At first sight it would seem that where we have a scattered population of plants almost always separated from one another by a foreign species, we should expect to find very little intraspecific competition. The poppy plants scattered about in a wheat field are not overcrowded, at least by one another. But there is serious overcrowding at a stage in the life-cycle where it can only be detected with the microscope, namely among the pollen-grains. If we examine the stigma of any flower after fertilisation we almost always find that there are far more pollen grains on it than are needed to fertilise all the ovules. We thus have to reckon with competition of two sorts. First, there is competition between different plants to pollinate their fellows (I am confining myself for the moment to plants where cross-fertilisation is the rule). Secondly, there is competition between pollen grains from the same parent.

The first type of competition evidently leads to the production of excessive amounts of pollen. No one who has walked through a pine-wood in summer, and above all no sufferer from hay fever, will doubt that more pollen is produced than is needed to assure that almost every ovule should be fertilised. But since a start of a few hours will probably ensure the success of a pollen grain in most cases, any plant which has pollen constantly available in fairly large amounts for fertilisation will be more heavily represented in the next generation that a more niggardly neighbour.

Secondly, there is competition between pollen grains of the same plant on the basis of the genes carried by them. This remarkable phenomenon was discovered by Heribert-Nilsson (1923), who called it certation. He studied the dominant type *rubrinervis* of

Oenothera Lamarckiana. Calling the gene responsible for its appearance R, the corresponding recessive *r*, he showed that a heterozygote R*r* fertilised with *r* pollen gave equal numbers of red- and white-veined plants. But the reciprocal cross, where the pollen grains consisted of equal numbers of R and *r*, gave 254 red and 93 white. Clearly the *r* pollen is severely handicapped in its competition with R. This is because the *r* pollen tubes grow more slowly. Although *rubrinervis* plants are far more cold-resistant than the normal, the gene R cannot establish itself in the species, because RR homozygotes are inviable, probably owing to linkage with a lethal. This is not an uncommon phenomenon. Several genes in Maize, including that for the waxy endosperm, handicap pollen tubes which carry them. For example, in plants with sugary endosperm Brink (1927) found that although pollen-grains with and without the gene causing waxy endosperm were formed in equal numbers, the former produced only 62 per cent. as many seeds as the latter.

Clearly a higher plant species is at the mercy of its pollen grains. A gene which greatly accelerates pollen tube growth will spread through a species even if it causes moderately disadvantageous changes in the adult plant. A gene producing changes which would be valuable in the adult will be unable to spread through a community if it slows down pollen tube growth. At one time I thought that such genes would be of overwhelming importance, owing to the great intensity of competition between pollen grains. This is incorrect, because if the competition is so intense that only one in n survives, the intensity of selection increases, not with n, but with $\log n$ or $\sqrt{\log n}$, *i.e.* very slowly (see p. 136).

While the behaviour of pollen grains depends to a considerable extent on the genes which they carry, this is fortunately not in general the case with spermatozoa, where Muller and Settles (1927) showed that the genes carried have, in *Drosophila* at least, no influence on their viability. The pollen grain represents the suppressed haploid generation of the higher plants, corresponding to the green generation of mosses, and has a physiology of its own, influenced by special genes. The spermatozoon, with its less distinguished past, does not depend in the same way on its nucleus. But similar influences undoubtedly come into play in higher animals. In the mouse a fair percentage, generally about a quarter, of the embryos die during pregnancy. There is not sufficient space or nourishment for them all, *i.e.* they compete with one another.

Hence in animals producing many young at a birth there will probably be selection in favour of rapid embryonic growth, and adult characters determined by genes causing rapid embryonic growth will spread through the species. We have here a possible cause for the orthogenetic evolution of unfavourable adult characters. This will tend to go on steadily because the prenatal environment is more constant than the adult environment. It is probably very significant that man and his immediate relatives only produce one child at a birth. For as Rubner (1908) and Bolk (1926) showed, the most deep-seated biological characteristic which distinguishes man from the other mammals is a marked slowing down of the rate of development.

I think it probable that competition based on embryonic growth-rates and possible gametic characters may account for many of the obscurer phenomena of evolution as disclosed by the geological record. Large size in an embryo necessitates the possession of circulatory and excretory systems, so that increased growth in the early stages of development implies the more rapid attainment of a certain degree of structural complexity. In other words certain genes will begin to act earlier, and if this is a general process, characters confined to the adult stage will be pushed back into the embryonic stage. Things may perhaps go further still, and embryonic characters be pushed back into the preceding life-cycle, thus explaining the "second childhood" of Ammonites. I have discussed this problem in detail elsewhere (Haldane, 1932a).

The converse process of neoteny, which causes the retention of larval or embryonic characters into the adult stage, is more likely where the larva or embryo is rather well off, and not subjected to intense competition. The larval Amblystoma is quite well equipped, and man is perhaps better protected during prenatal life than babyhood. As Madariaga (1925), put it, "A pregnant woman never leaves her infant behind." Nor does a pregnant monkey let it fall to the ground.

But it is in the struggle between adults of the same species that the biological effects of competition are probably most marked. It seems likely that they render the species as a whole less successful in coping with its environment. No doubt weaklings are weeded out, but so would they be in competiton with the environment. And the special adaptations favoured by interspecific competition divert a certain amount of energy from other functions, just as armaments, subsidies, and tariffs, the organs of international competition, absorb a proportion of the national wealth which many believe might be better employed.

If, like the authors of mediaeval bestiaries, I were using zoology to impart a moral lesson, I should suppress the paragraph which follows, and defend Kropotkin's point of view that intraspecific competition is always an evil, and mutual aid an important factor in evolution. The latter statement is clearly true, but it is also the case that some of the most striking cases of mutual aid appear to be useless, or at least of very small value, to one of the species which practises them, though necessary to its struggling individuals and admirable to man, who judges with a scale of values which is not merely biological.

One of the most striking of these cases is the insect-attracting flower. The bee or other fertilising insect usually finds in it a source of food, and is thus enabled to increase its numbers. The plant manages to secure the pollination of its own flowers and the carriage of its pollen to others. But both these can generally be secured quite efficiently by wind pollination, the former even by self-fertilisation. Self-fertilisation, as we shall see later, probably has its disadvantages, and wind-pollination doubtless requires a larger pollen production than does insect pollination. Nevertheless this large pollen production would involve less expenditure of material than the production of a corolla, scent, and nectaries. But a plant of an insect-pollinated species which did not attract insects, even if it secured fertilisation by accidental wind-blown pollen of other members of its species, would probably fail to fertilise any of its fellows, insects having carried pollen to them before any grains from it were blown to them. It would thus be inadequately represented in the next generation, and its genes would be eliminated by natural selection. Clearly this is not always so, for some plants have obviously abandoned insect fertilisation for wind pollination; but it must be so in general. It is a remarkable fact that the orchids, where, as Darwin showed, the adaptations for insect fertilisation are most strikingly developed, are not a very successful group. Sometimes even the insect does not benefit from the association, as in the case of the orchid *Cryptostylis leptochila*, which stimulates the male flies of the species *Lissopimpla semipunctata* which visit it to a biologically disadvantageous sexual activity (Coleman, 1928).

So with the bright colours and song of many bird species. They serve to attract the other sex, and incidentally delight humanity. But while they are probably preserved and enhanced by competition between members of the species, their value to the species as a whole is dubious.

Man is a dominant species, and is subject to the disadvantage entailed by that fact. The success or otherwise of a nation in the biological sense, *i.e.* the extent to which its members are represented in future generations, depends partly, no doubt, on the genes carried by it, partly on such accidents as the fact that England was in a favourable situation for colonising North America, and possessed large coalfields. But probably tradition in the broadest sense of the word has been an influence more important than either of these. Hence biological selection has largely been directed upon those characters which determine that one individual member of a nation shall be represented in the next generation by more children than another.

These characters include resistance to disease and a certain measure of physical vigour. But they do not include a number of the qualities which man himself finds most admirable, or which make for the multiplication of the species as a whole. Let me take two very different groups of men who have aroused the admiration of their fellows—the Christian saints and the winners of the Victoria Cross. Both include a large number who died young precisely on account of their heroic qualities. And the majority of saints were childless for other reasons. So with many of the great scientists and artists. Their choice of career made it economically or psychologically impossible for them to found families. Their genes are therefore unrepresented to-day, and their lives constituted a sacrifice of the future to the present.

The classes which are breeding most rapidly in most human societies to-day are the unskilled labourers. Society depends as much, or perhaps more, on the skilled manual workers, as on the members of the professional and ruling classes. But it could well spare many of the unskilled. There are, of course, tendencies acting in the opposite direction. Thus, on the whole, the earlier emigrants to new countries, whose descendants now constitute a considerable part of the population, were above the average in physical or psychological vigour.

However, I have no desire to discourse on eugenics; I merely wish to point out that some, at least, of the evils against which the eugenic movement is directed affect man not only because he is a social animal but because he is a dominant one. It can be shown mathematically that in general qualities which are valuable to society but usually shorten the lives of their individual possessors tend to be extinguished by natural selection in large societies unless these possess the type of reproductive specialisation found in social insects (see pp. 119–122). This goes a long way to account for the much completer subordination of the individual to society which characterises insect as compared to mammalian communities. Of course on Lamarckian principles one would expect exactly the opposite effect. The worker bees are descended from queens and drones, none of which have worked for very many generations, probably some million. One would expect the complex instincts of the worker to be gradually lost by disuse in these circumstances. They are not. Man on the other hand is, on the whole, induced by society to behave better than he would if left to his own devices. On Lamarckian principles he ought to be getting innately better in each generation. There is, unfortunately, no evidence for this view.

What is more, for reasons given in the appendix, I doubt if man contains many genes making for altruism of a general kind, though we do probably possess an innate predisposition for family life. But psychologists are perhaps right in regarding social life as an extension of family life, and theologians can use no more vivid metaphors than the fatherhood of God and the brotherhood of man. For in so far as it makes for the survival of one's descendants and near relations, altruistic behaviour is a kind of Darwinian fitness, and may be expected to spread as the result of natural selection.

But the altruism of the social insects is more thoroughgoing. That is why moralists tell us to imitate them. But it is hard to see how such behaviour could become congenitally fixed in a species which did not practise reproductive specialisation. It may be that, as Hudson (1919) suggested, man will adopt this practice, but the first steps to it, as pointed out by my wife (Haldane, 1926), would involve a very drastic interference with our present moral code, and would be most violently opposed by the moralists who would like us to imitate the insects in other respects.

A dominant species is perhaps subject to certain other disadvantages. It has large numbers and is consequently somewhat more variable, a fact first pointed out by Darwin, and later confirmed by Fisher and Ford (1929). The reason for the increased variability was first given by Fisher (see Appendix, pp. 117–118). He points out that the efficiency of selection is proportional to the variance of a species, so we may expect evolution to be relatively rapid in numerous species. On the other hand, their very numbers will tend to prevent them from breaking up into local races which are probably the precursors of new species. Of course such local races exist in dominant species, such as the herring, but a certain degree of crossing between them is almost inevitable unless they are isolated by geographical barriers. It is difficult to imagine that so widely distributed a fern as bracken (*Pteris aquilina*) is likely to break up into several species in England. On the other hand, the far rarer prickly buckler fern (*Nephrodium spinulosum*) has four British varieties which have been ranked as species, some of which have a rather restricted distribution. Clearly these varieties have a relative good chance of evolving along their own paths without being swamped by crossing.

I am speaking, of course, of species of which the members intercross freely. Things are quite different where self-fertilisation or apogamy is the rule. Here a common "species" consists of a swarm of innumerable varieties. *Taraxacum* and *Hieracium* are good examples. Of course the delimitation of species in such a case is quite arbitrary. Different pure lines, or Jordanons, of such species are adapted to slightly different environments, as Sukatschew (1928) showed, and for this reason the species as a whole is in equilibrium, different races living side by side in the same area, some perhaps thriving only in a rather restricted habitat, but not being swamped by the others, since they do not cross.

In such species we find the maximum of variability, and this enables them to fill a large number of slightly different ecological niches in the same territory. But this adaptation is probably less elastic than that of the outcrossing species. Such a species contains innumerable genes which under the existing circumstances are more or less disadvantageous. But if the environment alters, these genes will be able to form new combinations suited to the new environment. An out-breeding species is therefore far more elastic than one where cross-fertilisation is rare. If our climate were to change suddenly it is likely that most of our Jordanons of *Taraxacum* and *Hieracium* would perish. But the related genus, *Crepis*, which habitually outcrosses, would be in a position to produce new combinations of genes, some of which would survive.

We now begin to see some possible clues to the very different pictures of evolution given by palaeontologists and systematists. The former are mainly concerned with dominant species of aquatic animal, the latter mainly with relatively rare species, and as much with plants as with animals. Willis, for example, is a botanist. Now a widely distributed marine animal, which is mobile at least in one stage of its life cycle, and very rarely self-fertilising or apogamous, could rarely form a new species suddenly. A plant can often do so by polyploidy, whether as the result of hybridisation or otherwise. And in a rare plant or animal a local race has a far better chance of evolving without being swamped by hybridisation.

One entirely unsuspected type of natural selection has recently been shown by Fisher to be probable. Supposing a gene A constantly mutates to a, then if the species originally contains a gene B but its allelomorph b is equally viable in combination with A, while Aabb is more viable than AaBB, the effect of selection will be to substitute b for B. Fisher believes that the effect of this process has been to make most genes which frequently mutate recessive. Wright (1931) and I (Haldane, 1930a) have criticised this theory, and I doubt if it can stand in its original form. Nevertheless it undoubtedly has some truth in it, and there can be little doubt that mutation pressure has been a cause of evolution, if perhaps a less important one than Fisher believes.

We must now consider some of the other suggested causes of evolution. We have seen that in experiments lasting for a few score generations Lamarckian results are not generally obtained. When such results are claimed it is generally found that there has been conscious or unconscious selection. If we expose animals to a certain light, which alters their average degree of pigmentation, and then for several generations breed from those which have varied most in the direction favoured by the environment, we shall naturally find a permanent and inheritable change when we go back to the old environment. But so we do if we select those which have changed least. This was done by Bateson (Hall, 1928) with considerable effect in sugar beet. This plant commonly "bolts," *i.e.* puts up flowering shoots from a certain percentage of tubers in its first year. To eliminate the habit Bateson sowed his beet so early that about 50 per cent. bolted. The non-bolters were selected in this way over several years, and thus a strain was obtained which would not bolt under the ordinary conditions. He also selected a race which bolted more than usual, but did not, of course, interpret his result on Lamarckian lines.

One Lamarckian experiment at present stands out from the rest. Macdougall (1927, 1930) used a highly inbred stock of rats—in fact a very nearly pure line. These rats were placed in a tank from which they could escape by swimming. There were two landing places, one illuminated, one not so. Those that landed at the light received an electric shock. On landing in the dark they received no shock and were able to escape. The conditions were constantly altered so that it was useless always to swim ot the right or the the left. In the first few generations it took more than a hundred trials to establish the habit. This number was gradually reduced, and in the last ten generations fell from eighty to twenty-five, until after twenty-three generations only twenty-five trials were needed. Even when the slowest learners were selected to breed from, distinct progress occurred. The quality inherited was not an instinct for avoiding light, but apparently a cautious type of behaviour. Crossing trained and untrained stock gave a blend, even when the trained animal was a male; hence education of the young cannot explain the phenomenon. Now it is noteworthy that capacity to learn to find the way through a maze is not improved by many generations of training, as shown by Koltsova (1926), differences between races of rats remaining unaltered through ten generations. Several other experiments of this kind led to negative results. Thus Macdougall's results at present stand alone, and until they have been confirmed[1] it is rash to build a theory of evolution on them, especially as apparently equally striking results of the same character obtained by others in the past have not been substantiated by later workers. If they are correct we shall have to envisage the possibility that with the appearance of mind a new factor in evolution has come into being. But in such a case it will be extremely hard to explain why, for example, the instincts of worker bees do not come to resemble those of queens or drones.

Apart from Macdougall's work, the effects of use have not been shown to be inherited in the course of a few score generations. But it is suggested that they, or unknown internal cases which tend to make organisms vary in a definite direction, may have very large effects in times to be measured in thousands or millions of generations. I believe this apparently very plausible idea to be false, for the following reason. Variation does not appear to take place continuously, but by steps, even if they are very small. This must be so if it has a material basis at all. Now if the effect of the environment or of the unknown cause was to make a large proportion of the individuals of the race vary in each generation, we should expect to obtain measurable results within the period of an

---

[1] For a criticism see Sonneborn (1931).

ordinary experiment. If, on the other hand, only a few individuals change in each generation, we can show mathematically that the new character will not spread through the population in the face of a very mild degree of natural selection. Thus the most that these slowly acting causes of change could accomplish, would be the production of characters that were practically neutral as regards survival value. On the other hand, it is important to note that a recessive character, even if advantageous, has little chance of being selected unless it crops up fairly frequently as a result of mutation. Any single mutation will almost certainly disappear as the result of mere random extinction unless the polulation is highly inbred. Thus we certainly cannot neglect the frequency as a factor determining evolution.

But if we come to the conclusion that natural selection is probably the main cause of change in a population, we certainly need not go back completely to Darwin's point of view. In the first place, we have every reason to believe that new species may arise quite suddenly, sometimes by hybridisation, sometimes perhaps by other means. Such species do not arise, as Darwin thought, by natural selection. When they have arisen they must justify their existence before the tribunal of natural selection, but that is a very different matter. As Darlington (1928) pointed out, an allotetraploid hybrid usually possesses the vigour characteristic of hybrids, but without the usual disadvantages of hybrids—namely, either sterility or the failure to breed true. But this vigour is not the result of selection acting on random variations. It is the result of hybridisation.

Secondly, natural selection can only act on the variations available, and these are not, as Darwin thought, in every direction. In the first place, most mutations lead to a loss of complexity (*e.g.* substitution of leaves for tendrils in the pea and sweet pea) or reduction in the size of some organ (*e.g.* wings in *Drosophila*). This is probably the reason for the at first sight paradoxical fact that, as we shall see later, most evolutionary change has been degenerative. But further, as we saw in the last chapter, mutations only seem to occur along certain lines, which are very similar in closely related species, but differ in more distant species. This is fairly clear from the history of the domestic animals, where all sorts of mutations have been selected. The cow has shown a great capacity for variation in the direction of an increased milk yield, which has been exploited. The Scythians, according to Herodotus, lived largely on the milk of mares, and if mares had varied in the same way there can be little doubt that man would have selected mares with a high milk production, as he has selected she-goats. But we no more breed milch mares than racing bulls. So with other variations. Horns have appeared in the past in a great many races of hoofed animals. They do occasionally appear on horses, and it would very likely be possible to produce a race of horned horses. But in spite of the example of Pegasus, I doubt if the horse possesses the capacity for producing feathers.

We can now understand the parallelism and occasional convergence in evolution which has led many biologists to an anti-Darwinian standpoint. Related species will vary in similar directions and be subject to similar selective influences. They may therefore be expected to evolve in parallel. We need not be surprised if, say, the modern genus *Cervus* is descended from two distinct tertiary genera, *Cervavus* and *Dicroceras*, as many palaeontologists believe.

The believers in orthogenesis, due to internal causes, can still point to the parallelism in evolution of species which have developed similar characters, to all appearance useless or even harmful, as a prelude to extinction. Many such cases—for example, the development of large size or large horns—can, I think, be put down to the ill effects of competition between members of the same species. Others, such as the exaggerated coiling of *Gryphaea* (Chapter I) cannot at present be explained with any strong degree of likelihood.

But several explanations are possible. A study of the causes of death in man, animals, and plants leaves no doubt that one of the principal characters possessing survival value is immunity to disease. Unfortunately, this is not a very permanent acquisition, because the agents of disease also evolve, and on the whole more rapidly than their victims. Now, immunity is often correlated with physical characteristics. We all know the finely built type of man and woman who is most liable to succumb to phthisis. Macdonald (1911) produced strong evidence that immunity to measles and other diseases of childhood is correlated with hair and eye colour. In the United States the white and coloured populations die from very different types of infection. It seems likely that when a species is subjected to a series of attacks by an evolving parasite it may be forced along a path of structural change by its temporarily successful acquisitions of immunity. But in the end it may be driven, so to say, into a corner, where further immunity involves structural changes which are disastrous to it in its everyday life. Disease may have played a very important part in the decay of human civilisations. The same is possibly true of species, and especially of dominant species. Finally, Fisher, as the result of a rather intricate mathematical argument summarised on pp. 110–113, which is independent of the theory of dominance (in my opinion probably false), in terms of which it is stated, shows that, given certain assumptions, when a character such as size, determined by many genes, is being selected, the population is at any moment unstable if left to itself, and that the process of change will proceed further when selection stops or is reversed (see p. 113). Whether this momentum would carry a species to extinction is doubtful, but it might well carry it past the point of most perfect adaptation.

To sum up, it would seen that natural selection is the main cause of evolutionary change in species as a whole. But the actual steps by which individuals come to differ from their parents are due to causes other than selection, and in consequence evolution can only follow certain paths. These paths are determined by factors which we can only very dimly conjecture. Only a thorough-going study of variation will lighten our darkness. Although we have found reason to differ from Darwin on many points, it appears that he was commonly right when he thought for himself, but often wrong when he took the prevailing views of his time—on heredity, for example—for granted.

CHAPTER VI

Conclusion

"They saw the day how brief, the night how long,
The right how faint, how stark the groping wrong,
Man's lighted world how narrow, and how wide
The untrodden dark where all dark things abide;
With what grim toil the high gods keep at bay
The desperate leaguer of the haunts of day,
How at their side the souls of men outworn
Battle to hold the perilous pass of morn,
And, overborne, with agony maintain
The high adventure of the world, in vain."

Betts (1916).

We now come to the most difficult part of our task, the attempt to survey and evaluate evolution as a whole. As a preliminary it will be desirable to describe briefly the history of life on our planet.[1] Unfortunately, the opening acts of the drama are almost completely unknown to us. Geologists are too late a product of evolution to be able to tell us much of what went on before the Cambrian epoch, some 500 million years ago. One reason is that, until the Cambrian, scene-shifting for the drama of life was still in very vigorous progress. Almost all Pre-Cambrian rocks are severely folded, and the folding has blotted out most of the relics of life. We have fairly definite traces of the existence of calcareous seaweeds, of protozoa, and of worm-like marine animals with rudimentary legs. There is no evidence of life on land. Probably the intense folding confined the seas into smaller areas than at present, and the land surface of the globe was largely covered by almost rainless deserts. There were, however, at least two ice-ages before the Cambrian. The evidence of terrestrial rocks shows that the earth was already over a thousand

---

[1] Much the best account known to me of this history is given by Wells, Huxley, and Wells (1931) in *The Science of Life*, but I cannot share all their opinions on the causes of evolution, and note that they have misquoted me on this topic.

million years old in Cambrian times. It was quite probably nearly twice that age. We do not know how long life had existed on it, but probably for at least five hundred million years, possibly for much longer.

The first scene which is at all clear to us is the Cambrian epoch. Preserved in the Cambrian rocks we have remains of most of the main animal groups, or phyla. Thus we have skeletons of protozoa, sponges, coelenterates, worms, brachiopods, echinoderms, molluscs, and several branches of arthropods. The only important phylum unrepresented was the vertebrates, to which we belong. These did not appear till the Silurian, though presumably their boneless, wormlike ancestors were present in the Cambrian sea.

Most of the main animal types were thus already differentiated, though comparative anatomy, embryology, and biochemistry enable us to trace their relationships to one another, and to conjecture with some confidence the course of evolution of certain forms. Thus there can be little doubt that arthropods are descended from annelid worms, and that the various mollusca had a common ancestor. And a biochemist at least, who finds the same quite complex molecules in all plants and animals, can hardly doubt their common origin. There may be some reason in the chemical nature of things why all living creatures must contain glucose. But there appears to be no reason, other than common ancestry, why they should all contain dextrorotatory glucose, and none of them its mirror image.

Since the Cambrian we men can look back on a fairly steady progress among our ancestors. In the upper Silurian they were already fish with jaws (probably recently acquired, since jawless forms persisted beside them for a while). They rapidly developed paired fins from lateral folds. In the Devonian they developed bones, and probably about the end of that period left the water. The earliest land plants, somewhat resembling horsetails, date from the early Devonian. In the Carboniferous age our ancestors had definite legs, and were air-breathing swamp-dwellers not unlike the modern newt. Towards the end of the Carboniferous they were definitely reptiles and presumably capable of living on dry land, although incapable of lifting their bellies from the ground. By Permian times they were, however, probably walking on their feet, and beginning to differentiate their teeth. During the Mesozoic period, the age of the great reptiles, our ancestors were small animals whose remains have been inadequately preserved. They are classed as mammals, but we do not know at what stages they developed warm blood, a four-chambered heart, and mammary glands.

Meanwhile the land plants had been evolving rapidly. By the late Devonian many of the modern groups, such as ferns and club-mosses, were represented, and primitive seed-plants occur in the Lower Carboniferous, while by the end of the Coal Measures conifers were common. The dominant plants of the Mesozoic era resembled the modern Cycads, and possessed large flowers of a peculiar and rather primitive type. But they can hardly have been the ancestors of the modern flowering plants. The Caytoniales, an obscure middle Jurassic group, probably represent the first angio-sperms.

During the Cretaceous age, as the great reptiles disappeared, the mammals suddenly increased in size, and since then they have been the dominant land animals. The main changes in the mammals in the last thirty million years have been a general increase in brain size, and a specialisation in many directions, giving us such highly differentiated animals as the horse, stag, elephant, bat, and whale. But this last was largely accomplished in Eocene times. Thus whales and bats were already present at the end of the Eocene.

During the same period the modern flora took shape. Many modern families of flowering plants were already present in the upper Cretaceous, and there is very little evidence of serious evolutionary change in them during the tertiary period.

One mammalian order, the Primates, retained rather primitive limbs and teeth, but the brain developed to a considerable extent. Finally, a branch of them underwent an extraordinary arrest of development—foetalisation, as Bolk has termed it. The period of growth was greatly lengthened. There is a general law which gives the amount of food eaten by a mammal before it becomes adult. The same formula applies to the mouse and the elephant. For a man it must be multiplied by about seven. Putting the matter rather differently, a human baby doubles its birth weight in 180 days, a calf in 47 days. And yet the calf is a heavier animal, and is born in a more advanced state than a man, for both of which reasons we should expect it to develop more slowly.

During a very long period the higher vertebrates had been undergoing a process which de Beer (1930) called "clandestine evolution." The embryo had been diverging more and more from any past adult form. An embryo dog or chick has a relatively large brain, and the head is bent forwards so that its axis forms a right angle with that of the trunk. In later development this is straightened out. But in man the retardation of development, or of certain phases of it, has led to a retention of embryonic characters into adult life. The snout is a late development in other mammals. In man it never develops at all. Consequently we can look horizontally when standing erect, and focus both eyes on the same point. Many other human characters are similar to those of foetal or baby apes. Our cranial sutures do not close till the age of thirty; we do not develop bony brow ridges. We do not develop a second coat of hair. Some of the apes are intermediate in these respects. Thus the gorilla has a moderately human face, and is born with the head covered with hair, the rest of the body being covered later. A concise account of these facts, with references, is given by de Beer (1930).

Neanderthal man, as shown by his brow ridges and his more rapid development of teeth, was somewhat less foetalised than ourselves. He very probably reached maturity at a much earlier age than modern man, and was therefore probably less teachable. If human evolution is to continue along the same lines as in the past, it will probably involve a still greater prolongation of childhood and retardation of maturity. Some of the characters distinguishing adult man will be lost. It was not an embryologist or palaeontologist who said, "Except ye . . . become as little children, ye shall not enter into the kingdom of heaven."

The essential feature of the last stage of our evolution has thus been not so much the acquisition of new characters as the preservation of embryonic and infantile traits which had been developed at a period in the life cycle when the individual was sheltered from violence. Their retention by man has enabled him to shed a good deal of animalism.

The evidence regarding the last stages of human evolution is now accumulating fairly rapidly. The single specimen of *Pithecanthropus* from Java is now supplemented by several skeletons of the closely related *Sinanthropus* from China. These creatures had a skull capacity 40 to 50 per cent. larger than any ape, and only slightly less than those of the smallest brains of sane adult men. Their skeleton had many ape-like features. In view of their existence it is somewhat ridiculous to talk of the missing link. Their striking efficiency as links is shown by the fact that opponents of man's animal ancestry have not yet been able to decide among themselves whether they are to be regarded as the remains of apes or men!

Now this makes a very heartening story of fairly steady progress. A similar, though a far less detailed, history might be made out for a highly developed flowering plant such as a daisy or a snapdragon. But these histories of progress are exceptional. If we take some of the commonest early Palaeozoic animals we shall find that in general they have undergone no obvious progress. The lamp-shell *Lingula* has changed so little in 400 million years that the same generic name is used for the animals living to-day and in the Ordovician. The limpet *Patella* has persisted since the Silurian. The graptolites, ammonites, and trilobites changed along well-defined lines, but these changes were about as often in the direction of simplification as complication. Ultimately all three groups became extinct, probably without leaving descendants. One other line besides the vertebrates has shown marked progress. This is the insects. In the course of insect evolution since the Carboniferous the wings have been considerably improved, the mouth parts specialised, the eyes enlarged in many cases, the segments of the thorax fused together, and so on. But here there has been much degeneration. Many groups have lost their wings, some have become parasitic, mere sucking and egg-laying machines; and so on.

The usual course of evolution appears to have been a modification in the relative sizes and shapes of various structures, with very little real novelty. Occasionally true progress was made, as when insects and birds developed wings, but for every form which has improved, dozens have degenerated. Probably all the birds are derived from one ancestral species which took to the air, but very many have independently lost the power of flight. The ostriches and their allies, the dodo, the kiwi, the flightless parrots and rails of New Zealand, have all lost their flying power and gained nothing in exchange. Only the penguins have transformed their wings into fairly effective fins.

Very numerous groups whose ancestors were motile have taken to sessile habits or internal parasitism. Degeneration is a far commoner phenomenon than progress. It is less striking because a progressive type, such as the first bird, has left many different species as progeny, while degeneration often leads to extinction, and rarely to a wide-spread production of new forms. Just the same is true with plants. Many primitive forms have not progressed. A few have done so, but relapses of various kinds are equally common. Certainly the study of evolution does not point to any general tendency of a species to progress. The animal and plant community as a whole does show such a tendency, but this is because every now and then an evolutionary advance is rewarded by a very large increase in numbers, rather than because such advances are common. But if we consider any given evolutionary level we generally find one or two lines leading up to it, and dozens leading down.

I have been using such words as "progress," "advance," and "degeneration," as I think one must in such a discussion, but I am well aware that such terminology represents rather a tendency of man to pat himself on the back than any clear scientific thinking. The change from monkey to man might well seem a change for the worse to a monkey. But it might also seem so to an angel. The monkey is quite a satisfactory animal. Man of to-day is probably an extremely primitive and imperfect type of rational being. He is a worse animal than the monkey. His erect posture leads to all sorts of mechanical troubles, such as hernia and narrowing of the pelvis which makes childbirth painful and dangerous. The last stage in man's evolution certainly has its dark side. You will find a highly symbolic account of it in the second chapter of the Bible. Our first parents are represented as living in a state of ignorance, and then suddenly acquiring the knowledge of good and evil. This may conceivably be true. A decisive step from animal to human mentality may have occurred by mutation, though only a very convinced disciple of Harrison would nowadays ascribe such a mutation to changed diet. Perhaps it is more likely that it occurred in several steps. But this change is not chronicled as "Man's ascent to reason" or "Man's new nature" but "Man's shameful fall." The writer of Genesis very clearly felt "la honte de penser et l'horreur d'être homme," and we must remember that when we speak of progress in evolution we are already leaving the relatively firm ground of scientific objectivity for the shifting morass of human values. Nevertheless, just because we are men, we cannot avoid doing this, and we may as well attempt to do it as well as lies in our power.

Any such attempt involves us in philosophy, so before I make it I may as well frankly state my philosophical prejudices, for perhaps they should be rated no higher than that. My main prejudice is in favour of monism. Roughly speaking, the monistic systems may be grouped under absolute idealism, materialism, and intermediate systems such as the "neutral monism" of Russell. Materialism of course includes many forms far more subtle than the crude materialism of fifty years ago, and if you are willing to concede enough unexpected properties to so-called dead matter it becomes distinctly idealistic. To quote Lenin's words, "For every materialist the laws of thought that reflect the forms of the real existence of things are totally like, and in no way different from those forms." If Lenin was right, it seems to me not unlikely, so much the better for "things."

Over against these stand various pluralistic systems which hold that the distinction between different minds, or between mind and matter, is irreducible. My objection to them is just that they proclaim certain problems to be insoluble merely because three thousand years of thought by a few members of a species which may have many thousand million years ahead of it has not yet solved them. For a scientific man a philosophy is a programme rather than a creed. Some parts of the monistic programme may be impossible, but we need not abandon it until a really serious attempt has been made to carry it out. Thus a study of cerebral physiology is leading to results which at present can be interpreted either as the mind-like nature of certain objects which we generally call material systems, or as the mechanical character of conscious behaviour. Until the attempt has failed we need not, I think, fall back on mind–body dualism. Meanwhile, monism has the advantage that if it is wrong it will ultimately lead to self-contradiction, whereas dualistic systems, which purport to give a less complete account of the world, are therefore less susceptible of disproof. My preference among monistic systems has been stated elsewhere, and is irrelevant to the present discussion.

Particularly hostile to true scientific progress are the extremer forms of the doctrine of emergence. According to these, a material system of a certain degree of complexity suddenly exhibits qualitatively new properties such as life or mind, which cannot be explained by those of the constituents of the system. There is clearly an element of truth in this view. We can only discern a little mind in a dog, and at present none in any oyster or an oak. Nevertheless science is committed to the attempt to unify human experience by explaining the complex in terms of the simple. This may be a vain endeavour, but I do not at present see any evidence of its vanity.

I will give an example of its success in the realm of physics. J. J. Thomson and Rutherford showed that the hydrogen atom could be separated into two components, the electron and the proton, which behaved in many situations like very small spherical electric charges attracting one another according the Coulomb's law. But the hydrogen atom has very complex properties. It emits a series of characteristic radiations whose frequencies are related by definite laws like those of the notes of a piano. This is the simplest example of emergence, or holism, the properties of the whole being far more complex than those of the parts. It held up the progress of theoretical physics for a generation. Then de Broglie (1930) produced wave mechanics. To explain these facts, he said, we must attribute to the electron certain undulatory properties. These properties were soon afterwards experimentally verified by G. P. Thomson (1930) and others. The electron and the proton were shown to be more complex than they at first appeared, though by no means so complex as the hydrogen atom.

I regard this as a model for scientific explanation. If we ever explain life and mind in terms of atoms, I think we shall have to attribute to the atoms the same nature as that of minds or constituents of mind such as sensations. Many philosophers have, of course, said this long ago, but in my opinion the details of all metaphysical systems have been incompatible with certain observed facts. Moreover, these systems have generally been used to support some particular form of religion or irreligion adopted by their framers on grounds which were largely sentimental or economic.

Only when science has progressed to this stage will we be able, so it seems to me, to speak with any great confidence about the mind-like qualities, if any, to be attributed to the universe as a whole. Such are the philosophical prejudices with which I look at evolution.

I have given my reasons for thinking that we can probably explain evolution in terms of the capacity for variation of individual organisms, and the selection exercised on them by their environment. This excludes the action of a mind or minds higher than that of the evolving individuals, except in so far as such a mind is concerned in the general nature of the universe and its laws, a question too vast to discuss here.

The most obvious alternative to this view is to hold that evolution has throughout been guided by divine power. There are two objections to this hypothesis. Most lines of descent end in extinction, and commonly the end is reached by a number of different lines evolving in parallel. This does not suggest the work of an intelligent designer, still less of an almighty one. But the moral objection is perhaps more serious. A very large number of originally free-living Crustacea, worms, and so on, have evolved into parasites. In so doing they have lost to a greater or less extent, their legs, eyes, and brains, and have become in many cases the source of considerable and prolonged pain to other animals and to man. If we are going to take an ethical point of view at all (and we must do so when discussing theological questions), we are, I think, bound to place this loss of faculties coupled with increased infliction of suffering in the same class as moral breakdown in a human being, which can often be traced to genetical causes. To put the matter in a more concrete way, Blake expressed some doubt as to whether God had made the tiger. But the tiger is in many ways an admirable animal. We have now to ask whether God made the tape-worm. And it is questionable whether an affirmative answer fits in either with what we know about the process of evolution or what many of us believe about the moral perfection of God.

We can answer the question in three ways. We can regard the dark as well as the bright side of evolution as a manifestation of divine ingenuity. "I make peace, and create evil: I the Lord do all these things" (*Isaiah*). Secondly, we can go for our answer to Plato. Socrates in the "Republic" says, "God therefore, since He is good, cannot be responsible for all things, as the many say, but only for good things." This answer, however, leads us into Manichaeanism, for the tapeworm presents just as much ingenuity in construction (if we regard it as designed) as does the rose. We should have to give the Devil credit for a large share in evolution. Or, finally, we can say that at present it does not seem necessary to postulate divine or diabolical intervention in the course of the evolutionary process. The question whether we can draw theological conclusions from the fact that the universe is such that evolution has occurred in it is quite different, and very interesting.

The minds of evolving animals are, however, concerned in evolution in several ways. If Macdougall is confirmed, they are concerned in it directly. In any case it is important to note with Elton (1930) that every animal has at least a motile period in its life cycle during which it chooses its environment, and therefore the system of selective agencies to which it will be subjected. Moreover, in the cases of sexual selection and the evolution of flowers, survival value has been determined by animal aesthetics, which are not altogether unlike our own. In social animals the form of the society, and therefore the type of selection, depends on the social instincts of the individual animals. These facts would be irrelevant if we regarded mind simply as a product, and a by-product at that, of natural selection acting on random variations. But I do not think we can do so. Clearly we cannot if we adopt an idealistic standpoint, but I do not think such a view is consistent even with materialism, as I propose to show.

For the materialist mind is a by-product or epiphenomenon of certain material systems. These systems are very complex and easily deranged. Now, in biochemistry we find plenty of examples of material systems which have very complex and specific properties. For example, we have the oxygen-carrying pigments of blood, which must be able to take up and unload oxygen very rapidly over quite a small range of gas pressures. Only two types of pigment, haemoglobin and haemocyanin, are of any great value in this respect. Similarly it is to be expected that the types of material system associated with mind, and hence the types of mind possible, will be severely restricted. We shall not be surprised to find considerable similarities between minds which have developed on quite different lines of descent. We shall not regard it as a mere coincidence that man cannot merely sympathise with the bee's devotion to its hive, but with its preferences regarding the colour and smell of flowers, and with its habit of dancing when it has satisfied its desires. Clearly, if we are idealists, these resemblances will be still more easily intelligible.

For such reasons as these I do not share the view that mind, as we know it, cannot be a product of evolution. An essential element of evolution is variation. Variation is at random in the sense that it may lead in many directions, mostly of no survival value, and that those which possess survival value for the individual may lead to degeneration and extinction of the species. But it follows chemical and biological laws, and only certain combinations will lead to mind. If we are to have mind at all, it must probably conform to certain laws. There is no need to suppose that these laws, any more than those of biochemistry, are products of natural selection. Selection no doubt accounts for certain details, but in all probability not for the general character of mind.

At this point or earlier some of my biological readers will doubtless object that it is unscientific to describe animal and human behaviour in terms of mind. We should always try to explain it (they will say) on physico-chemical lines. This objection seems to me to savour of philosophy rather than science. As a scientist I am engaged in an attempt to unify my experience, and will describe A in terms of B, or B in terms of A, as it suits my convenience. The idealist wants me always to describe matter in terms of mind, the materialist makes the opposite demand. Now in plane geometry I use point co-ordinates or line co-ordinates as it suits me. Although on the whole the point is the simpler idea, it may suit my convenience to describe every point by specifying two lines which meet in it. The idealist, to speak metaphorically, would like me always to do this; the materialists would forbid it. Personally I find geometry difficult enough to excuse my employing any co-ordinate system I choose. So with biology. It is only in systematic philosophy or mathematics that we can as yet attempt to deduce a complex system from a few premises. The bulk of science is still in the heuristic stage.

Now the hypothesis that mind has played very little part in evolution horrifies some people. Shaw's preface to "Back to Methuselah" is a good example of a strong emotional reaction. He admits that Darwinism cannot be disproved, but goes on to state that no decent-minded person can believe in it. This is the attitude of mind of the persecutor rather than the discoverer. Shaw's case is complicated by his admiration for Samuel Butler, who was undoubtedly a better stylist than Charles Darwin. But he had less respect for facts.

My reaction is entirely different. If evolution, guided by mind for a thousand million years, had only got as far as man, the outlook for the future would not be very bright. We could expect very slow progress at best. But if now for the first time the possibility has arisen of mind taking charge of the process, things are more hopeful. We certainly do not know enough at present to guide our own evolution, but we have only been accumulating the knowledge necessary for such guidance during a single generation. There is at least a hope that in the next few thousand years the speed of evolution may be vastly increased, and its methods made less brutal. If human evolution continued in the same direction as in the immediate past, the superman of the future would develop more slowly than we, and be teachable for longer. He would retain in maturity some characteristics which most of us lose in childhood. Certain shades of the prison house would never close about him. He would probably be more intelligent than we, but distinctly less staid and solemn.

Various imaginative writers have attempted to depict such supermen. Wells' (1923) "Men like Gods" are probably no better than the best thousandth[1] of the present human race placed in a

[1] Wells puts the proportion at half. I do not share his high opinion of his fellows.

favourable environment. Shaw's (1921) ancients in "Back to Meth-uselah" have reversed the most essential step by which man evolved from monkeys. They reach complete maturity in about four years, and then lose most of the characters which we find attractive in our fellow-creatures. To a biologist they are unconvincing. On the other hand, Stapledon (1930) in "Last and First Men" describes the human race 2×10^9 years hence. His "last men" require two thousand years to come to maturity, and although they have five eyes and other evidences of evolutionary change, besides great intellectual and moral perfection, are likeable creatures who fall in love, indulge in sport and ritual, and enjoy life like ourselves, only more so. Fortunately the account of their origin and nature is much more consonant with what we know of biology than is that of Shaw's creations. If anyone desires a speculative, but not (in the light of our present knowledge) wildly impossible, account of man's future, I advise them to read "Last and First Men." Wells (1895) and I (1927d) have given less alluring accounts, both involving a bifur-cation of the human species into two, each of which loses certain qualities which we admire in contemporary man.

Bergson attributed evolution to an *élan vital*, or vital impulse, which pushed organisms forward along the path of evolution. He laid special stress on convergence, *i.e.* the production of very similar structures by different means in different lines of descent. For example, he pointed out that vertebrates and molluscs have inde-pendently developed eyes with a lens and retina, and regarded this as disproving Darwinism. Now, as far as we can see, there are only four possible types of eye, if we define an eye as an organ in which light from one direction stimulates one nerve fibre. There is the insect type of eye, a bundle of tubes pointing in different directions, and three types analogous to three well-known instruments, the pinhole camera, the ordinary camera with a lens, and the reflecting telescope. A straightforward series of small steps leads through the pinhole type to that with a lens, and it is quite easy to understand how this should have been evolved several times. On the other hand, the type with a reflector would be little use in its early stages, and has never been evolved. However, if I were designing an animal as a construct with no historical background, like the ideal state, I should very probably give it an eye with a concave mirror rather than a lens.

But the main objection to *élan vital* is that it is so very erratically distributed. That sturdy little creature, the limpet, has watched the legions of evolution thunder by for some three hundred million years without changing its shell form to any serious extent. And the usual course taken by an evolving line has been one of degeneration. It seems to me altogether probable that man will take this course unless he takes conscious control of his evolution within the next few thousand years. It may very well be that mind, at our level, is not adequate for such a task, probably on account of its emotional rather than intellectual deficiencies. If that is the case we are perhaps the rather sorry climax of evolution, and less can be said in favour of existence than many of us suppose.

If I were compelled to give my own appreciation of the evolutionary process as seen in a great group such as the Ammonites, where it is completed, I should say this: In the first place, it is very beautiful. In that beauty there is an element of tragedy. On the human time-scale the life of a plant or animal species appears as the endless repetition of an almost identical theme. On the time-scale of geology we recapture that element of uniqueness, of *Einemaligkeit*, which makes the transitoriness of human life into a tragedy. In an evolutionary line rising from simplicity to complexity, then often falling back to an apparently primitive condition before its end, we perceive an artistic unity similar to that of a fugue, or the life work of· a painter of great and versatile genius like Picasso, who began with severe line drawing, passed through cubism, and is now, in the intervals between still more bizarre experiments, painting somewhat in the manner of Ingres. Possibly such artistic work gives us a good insight into the nature of the reality around us as any other human activity. To me at least the beauty of evolution is far more striking than its purpose.

In my moments of wilder speculation I sometimes go further. I imagine that associated with an evolving line there may be some "emergent," just as mind is associated with brain. Royce (1901) tried to give a concrete picture of such an emergent as a mind with a vast time-scale, and suggested that the intense feelings associated with reproduction were in that mind as well as our own. If there is an element of truth in such speculation, I question whether such an emergent should be regarded as probably mind-like. Man, if it is anything more than an aggregate, is presumably no more like an individual man than the British nation is like the lady on the reverse of a penny. We have already seen reasons to doubt whether mind has played any important part in guiding evolution, nor should I expect it to appear in the absence of brain. My suspicion of some unknown type of being associated with evolution is my tribute to its beauty, and to that inexhaustible queerness which is the main characteristic of the universe that has impressed itself on my mind during twenty-five years of scientific work.

But I realise only too well how futile must be any attempt to pass judgment of value on evolution until we know more about it. The first five chapters of this book have served, I hope, to reveal the depths of our ignorance. But they do also reveal the fact that our ignorance is diminishing. We can say appreciably more about evolution to-day than was possible ten years ago. The way to still further knowledge lies largely in the accumulation of more facts concerning variation and selection. But man is a theorising animal. He is continually engaged in veiling the austerely beautiful outline of reality under myths and fancies of his own device. The truly scientific attitude, which no scientist can constantly preserve, is a passionate attachment to reality as such, whether it be bright or dark, mysterious or intelligible. I would have you remember of this book only so much as I have been able to show you of the real, and forget the framework of speculation which, like myself, is transitory and ephemeral.

Appendix: Outline of the Mathematical Theory on Natural Selection

INTRODUCTION

The earliest work on this topic was done by Pearson and his colleagues. They accumulated numerous data on the inheritance of stature and other continuously varying characters in man. These enable a perfectly definite answer to be given to certain questions. *E.g.* "If, in the group of the English population considered, all parents either of whose heights was less than 5 ft. 6 in. had been prevented from breeding, what would have been the mean height of the offspring of the remainder?" For we have only to eliminate these parents from the tables, and make the ncessary calculations. But when it was attempted to extend the method to more remote generations, the results were less satisfactory. For the coefficients of correlation between an individual and his remoter ancestors, *e.g.* great-grandparents, cannot be obtained directly from the data, but are calculated indirectly. And the calculation rests on the particular theory of genetics held by Pearson. The results of such calculations (*e.g.* Pearson, 1930) are not in harmony with experimental results obtained in other organisms. Nevertheless Pearson's observations remain as fundamental data for future work.

The theory of selection in Mendelian populations is mainly due to R. A. Fisher, S. Wright, and myself. Fisher's work is largely collected in his book,* and for that reason I shall give only a summary account of it. My own is published in the *Transactions* and *Proceedings of the Cambridge Philosophical Society*, and reprints of some papers are no longer available. Other important papers are those of Kemp, Warren, and Norton.

THE MEASUREMENT OF THE INTENSITY OF SELECTION

The simplest case occurs when the population consists of only two types which do not interbreed, and when generations do not

---

* *The Genetical Theory of Natural Selection.*

overlap, as in annual plants. Suppose that for every n offspring of type A in the subsequent generation, type B gives, on an average $(1 - k)n$, we call k the coefficient of selection in favour of A. Unless k is very small it is better to use the difference of Fisher's "Malthusian parameters" for the two types. If type B gives $e^{-\kappa}n$ offspring, κ is the difference in question. When both are small, k and κ are clearly almost equal; but in general κ may have any real value, while $k = 1 - e^{-\kappa}$, or $\kappa = -\log_e (1 - k)$, so that k cannot exceed 1, but may have any negative value. Note that k in general depends both on birth-rates and death-rates. Thus in England unskilled workers have a higher death-rate, both adult and infantile, than skilled, but this is more than balanced by their greater fertility. In calculating k we may make our count at any period in the life-cycle, provided this period is the same in both generations. It is often convenient to call $1 - k$ the relative fitness of type B as compared to A.

We can generalise k in a number of ways. In general the types A and B will interbreed. If we are dealing with a difference due to a plasmon (see p. 21) this makes no difference to our calculation. Otherwise the most satisfactory method of defining the intensity of selection would be as follows. Individuals should be counted at the moment of fertilisation. Then if for every n children of an A ♀ , a B ♀ has $n(1 - k_1)$, and $n(1 - k_2)$ is the similar figure for a B ♂ , we have the two coefficients k_1 and k_2. The same calculation holds for an hermaphrodite species. If selection operates mainly through death-rates, k_1 and k_2 are likely to be nearly equal. If it operates mainly through fertility this is not so. For example, male sterility is quite common in hermaphrodite plants, and in the male-sterile group k_2 will be 1, while k_1 is small. Where (as in heterostylic plants) the success of a mating depends on the precise combination of parents, special methods must be used.

Where generations overlap, as in man, a specification of the intensity of selection involves the use of definite integrals. I showed (Haldane, 1927a) that if the chance of a female zygote (whether alive or dead) producing offspring between the ages x and $x + \delta x$ is $K(x)\delta x$ for type A, $[K(x) - k(x)]\delta x$ for type B, then selection proceeds as if generations did not overlap, the interval between generations being

$$\left[\int_0^\infty xK(x)dx\right]\bigg/\left[\int_0^\infty K(x)dx\right],$$

and the coefficient selection

$$k = \int_0^\infty k(x)dx.$$

This is true provided k is small. Similar conditions hold when selection operates on both sexes, and Norton (1928) has discussed the rather intricate mathematical problems arising in this case. In what follows we shall confine ourselves to the case where generations do not overlap, as this involves very little loss of generality, and greatly simplifies the mathematics. When generations overlap, the finite difference equations which will be developed later become integral equations.

The values of k vary greatly. For a lethal recessive $k = 1$, $\kappa = +\infty$; for many of the semi-lethal genes in *Drosophila* k exceeds 0.9, and probably for most of those studied 0.1. In many cases, *e.g.* those characterising the races of *Taraxacum* described on pp. 48–50, it is fairly large, and positive or negative according to the environment. For *Primula sinensis* we only possess data regarding mortality, as opposed to fertility. Here k varies from less than 0.01 upwards. Of 24 mutant genes 20 are neutral or nearly so, two give values of k about 0.05, one about 0.10, and one about 0.6. For some of the colour genes in mice it appears to be less than 0.05, while for the genes determining banding in *Cepea* it is 10^{-5} or less.

CAUSES INFLUENCING THE INTENSITY OF SELECTION

In general this problem is too complex for mathematical treatment, but two cases have been discussed. Fisher (1930) discusses the selective value of simultaneous changes in the various parameters of an organ, *e.g.* eye length and corneal curvature (see p. 57). To simplify matters, consider first variation involving only two parameters x and y, whose optimal values are a and b, in a given environment, and with the other characters of the organism constant. Then we can represent any organism by a point in the x, y plane. Clearly the farther away we go from (a, b) in any direction the worse off is the organism. The points representing organisms of equal viability will lie in closed curves round (a, b). Fisher considers the special case where these curves are circles. They will in general be ellipses, whatever the scales of x and y, because a harmonious variation of the two will be less unfavourable than an inharmonious one. They can only be converted into circles by choosing very artificial characters for our x and y. For example, an increase in pigmentation in an animal might be disadvantageous unless balanced by an increase in the capacity of its liver for storing vitamin D during sunny weather. But it would be very artificial to take the sum and difference of two numbers representing pigmentation and storing power, as our variables, rather than the numbers themselves.

If instead of only two variables, x and y, we have n variables, we shall have closed "varieties," *i.e.* hypersurfaces in n dimensions round our optimum point. Consider now a change in the variables x, y, etc., which is represented by a motion of the representative point through a distance r. What is the chance that the new representative point will lie inside the "surface" on which the original point lay? If it is inside, the organism represented will be fitter. If r is very small, this chance is $\frac{1}{2}$, for in general the curvature of the "surface" within a small distance r is negligible. When r is larger than the distance to the farthest point on the "surface," the chance is 0. Fisher shows that in the special case of a hyper-sphere in a large number n of dimensions, of diameter d, this probability approximates to

$$p = \frac{1}{\sqrt{2\pi}} \int_x^\infty e^{-\frac{1}{2}u^2} \, du,$$

where $x = (r\sqrt{n})/d$. When r is equal to the radius of the sphere, *i.e.* a change in the right direction would achieve the best possible results, $p = 1/(\sqrt{n\pi/2} \, e^{n/4})$, approximately.[1] This quantity is very small when n is large, but it is never zero. Provided, however, the change proceeds by small steps, represented on the diagram by distances less than $n^{-\frac{1}{4}}d$, the probability of an ultimate adaptation is large. Such a course of events is sure but slow.

I (Haldane, 1931a) have considered the effect on selection intensity of varying the intensity of competition. A simple example from artificial selection will make the matter clear. Consider two of Johannssen's pure lines of beans, line A with a mean weight of $a + \lambda$ and standard deviation $\sigma + \mu$, and line B with mean weight $a - \lambda$ and standard deviation $\sigma - \mu$. In actual fact λ and μ are often fairly small compared to σ, *i.e.* the distributions are nearly the same. Now suppose we start with a mixture of equal large numbers from the two pure lines, and choose all beans whose weight exceeds $a + x$, what

---

[1] See Notes by E. G. Leigh Jr. on p. 210 *et seq.*

will be the proportion of B to A in the chosen batch? We can call this ratio $1 - k$, and specify the intensity of competition by z, the proportion of beans eliminated to chosen. If competition is no more intense than between children in a civilised state where the infantile mortality is less than 9 per cent., z is less than 0.1. If it is as intense as among the pollen grains of a *Sequoia gigantea*, of which one in the course of some centuries fertilises a seed that grows into an adult "big tree," z would be very large, perhaps exceeding 10^{12}. Values from 100 to 10,000 are common in nature.

Common sense tells us that k increases with z, but I think exaggerates the rate of increase. If λ and μ are sufficiently small compared to σ, we have

$$\frac{1}{z + 1} = \frac{1}{\sqrt{2\pi}} \int_{x/\sigma}^{\infty} e^{-\frac{1}{2}u^2} du$$

and

$$k = \frac{2(\lambda\sigma + \mu x)(z + 1)}{\sqrt{2\pi\sigma}} e^{-x^2/2\sigma^2}$$

the value of x being obtained from the first equation.[2] If $\mu = 0$, *i.e.* the standard deviations are the same, the relationship between k and z is as shown in Fig. 9, where $q = (k\sigma)/(2\lambda)$. When z is large, $k = (2\lambda/\sigma) \sqrt{\log_e (z^2/2\pi)}$ approximately. *I.e.* the intensity of selection only increases very slowly indeed with z. Thus k only increases 9 times when z is raised from 1 (50 per cent. elimination) to 10^{12}. If μ is not zero, *i.e.* the standard deviations of the two populations are not the same, the value of k ultimately becomes proportional to log z. But in this case selection always changes sign at some value of z. It is easy to see why this should be so. If we were only selecting one bean in a million, we should probably favour, not the race with the higher average weight, but that with the higher spread of weights, *i.e.* the more variable race. So *intense competition favours variable response to the environment rather than high average response* (see pp. 64, 65). Were this not so, I expect that the world would be much duller than is actually the case. If the coefficient of variation is the same in the two populations, *i.e.* $\lambda/a = \mu/\sigma$, the value of k changes sign when $z = 0.1886$, *i.e.* with a mortality of 15.9 per cent..

---

[2] See Notes by E. G. Leigh Jr. on p. 210 *et seq.*

How far can these calculations be applied to natural selection? In some cases the analogy is very close. Thus pollen tubes carrying different genes grow at rates which can be expressed by frequency curves. The first to arrive at ovules produce seeds, the others die. In this case theory and experiment agree well. The same may be true for spermatozoa. In the case of seedling plants selection is intense, and those which escape an early death generally produce a fair amount of seed. The character selected is probably rapid growth-rate. On the other hand in many higher animal species matters are different. Selection is not very intense during childhood, and depends to a large extent on effective fertility during adult life. Here

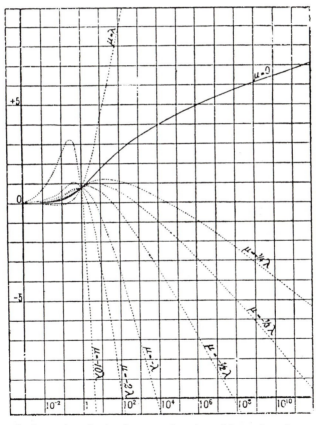

Figure 9. Intensity of selection as a function of intensity of competition. Abscissa, $\log_{10} z$; ordinate, $q = k\sigma/2\lambda$.

we cannot give a definite value to the number z, and further investigation is required.

To sum up, a change in the intensity of selection may reverse the relative fitness of two types, and it is not always true that intense competition means intense selection.

SLOW SELECTION FOR A FULLY DOMINANT GENE IN A LARGE POPULATION

I have dealt with this case rather fully (Haldane, 1924a). If the population is apogamous or self-fertilising, or practises obligatory brother–sister mating, the situation is the same as if the unit inherited is a plasmon. If u_n be the ratio of A to B in the nth generation, then $u_{n+1} = u_n/(1 - k)$, or $e^k u_n$, since k is small.

$$\therefore \quad u_n = e^{kn} u_o, \quad \text{or } kn = \log_e \frac{u_n}{u_o} \qquad (1)$$

Thus the ratio u_n increases in a geometrical progression. If we plot its logarithm against the number of generations we get a straight line (Fig. 8). No other system of inheritance or mating is more effective than this in promoting selection. *The number of generations required for a given change in the population is inversely proportional to the intensity of selection. This is true for all systems of slow selection.*

Now consider a group mating at random. In what follows we shall constantly use the variable u_n to denote the ratio of dominant to recessive *genes* in the nth generation. It can be shown that a change in the system of mating does not affect the value of u_n, which is only altered by selection. If [AA] denotes the number of AA zygotes, and so on, it is clear that

$$u_n = \frac{[AA] + \frac{1}{2}[Aa]}{\frac{1}{2}[Aa] + [aa]}$$

In a random-mating group a population composed of the three genotypes in the ratio $u^2 AA : 2uAa : 1aa$ is stable in the absence of selection, and any group whatever reaches this stable equilibrium after a single generation of random mating. This is only true for autosomal genes. For a sex-linked character the stable population is $[u^2 AA : 2uAa : 1aa] \female$ and $[uA : 1a] \male$ if the female is homogametic. Equilibrium is not reached at once, but the approach is rapid, and in each generation the difference between the values of $1/(1 + u_n)$ in the actual and final populations is halved. If selection is slow, not only is the population always *exactly* in equilibrium for autosomal gene-pairs, but nearly so for sex-linked.

Now after selection the population $u_n^2 AA : 2u_n Aa : 1aa$ is reduced to $u_n^2 AA : 2u_n Aa : (1 - k)aa$

$$\therefore \quad u_{n+1} = \frac{u_n(u_n + 1)}{u_n + 1 - k} \tag{2}$$

For the moment we are only considering the case when k is small.

$$\Delta u_n = u_{n+1} - u_n = \frac{ku_n}{u_n + 1 - k} \tag{3}$$

If k is small we can neglect it in comparison with 1, and treat the above as a differential equation, *i.e.* write

$$\frac{du}{dn} = \frac{ku}{1 + u}$$

Hence

$$kn = u_n - u_o + \log_e \left(\frac{u_n}{u_o}\right) \tag{4}$$

or if $u_o = 1$, $kn = u_n + \log_e u_n - 1$.

The actual proportion of recessives is $z_n = 1/(1 + u_n)^2$.

So $kn = z_n^{-\frac{1}{2}} + \log_e (z_n^{-\frac{1}{2}} - 1) - 2$. The ratio of dominants to recessives $p_n = u^2 + 2u$.

So

$$kn = \sqrt{1 + p_n} + \log_e (\sqrt{1 + p_n} - 1) - 2.$$

In Fig. 8 $\log p_n$ is plotted against kn. So long as p_n is small, *i.e.* dominants few, p_n increases or decreases in geometrical progression. When p_n is large, *i.e.* recessives few, p_n is roughly equal to $k^2 n^2$, so it increases or decreases pretty slowly, the proportion of recessives being $1/(k^2 n^2)$. In other words, *selection is not very effective on populations containing only a small proportion of recessives*. In future we shall speak of selection as fairly rapid whenever the successive values of u_n approximate to a geometrical series.

A number of other types of selection in random-mating groups have been investigated. Confining ourselves to autosomal genes, we may consider the case where selection operates slowly on the two sexes with intensities k_1 and k_2. This leads to changes at the same rate as a selection of intensity $\frac{1}{2}(k_1 + k_2)$ acting on both equally. We can consider a selection which operates only between members of the same family. Such would be selection operating entirely on embryonic characters, where the number of survivors is limited, and is not increased by the possession of the character in question. Here the march of selection is given by equation (4). But the value of k must be multiplied by $\frac{1}{2}$ if all members of a competing family have both parents in common, by $\frac{3}{4}$ if they have a common mother, but several different fathers.

Where selection operates on the gametes of one gender, $e.g.$ pollen tubes, we have $kn = 2 \log_e u_n$, or $u_n = e^{\frac{1}{2}kn}$. The proportion of recessives is $y_n = 1/(1 + e^{\frac{1}{2}kn})^2$, so when recessives are few they increase or decrease in geometrical progression, as do dominants when these are few. Hence selection of this type will begin operating at once on a new and therefore rare recessive gene. The fact that the expression of this gene in the diploid phrase is disadvantaged will not begin to stop its spread until it is fairly common. If k be the coefficient of selection in the diploid, equilibrium is reached when $u = (2k/1) - 1$. If competition is only between gametes from the same individual, as is commonly the case, the rate of selection is halved, but the phenomenon is qualitatively similar.[3]

In the case of a sex-linked gene, if the coefficients of selection against the recessive types are k_1 in the homogametic (usually female) sex, and k_2 in the heterogametic, the population is nearly (but not quite) in equilibrium apart from the effects of selection, and[4]

$$u_{n+1} - u_n = u_n \left(\frac{2k_1}{u_n + 1} + k_2 \right)$$

so that

$$(2k_1 + k_2)n = \log_e \left(\frac{u_n}{u_o} \right) + \frac{2k_1}{k_2} \log_e \frac{u_n + 1 + \dfrac{2k_1}{k_2}}{u_o + 1 + \dfrac{2_1}{k_2}}$$

When $k_2 = 0$ this becomes

$$\tfrac{2}{3}k_1 n = u_n - u_o + \log_e \frac{u_n}{u_o}.$$

[3, 4] See Notes by E. G. Leigh Jr. on p. 210 $et\ seq.$

If however k_2 is not 0, *i.e.* selection is at all effective on the heterogametic sex, it proceeds at a reasonable rate even when recessives are rare. It is worthy of note that in species with haploid males, *e.g.* the social hymenoptera, all genes behave as sex-linked. This fact may have accelerated their evolution.

A summary of the numerical results deducible from these equations is given in Table VI. In each case it is supposed that a dominant gene is favoured, the intensity of selection being given by $k = 0.001$. The number of generations required for a given change is tabulated. The second column gives the sex on which selection acts, in the case of a sex-linked gene, and the type of selection otherwise. Only the ordinary type of selection is considered as regards sex-linked genes. The third column gives the sex to which the figures refer. Exactly the same figures would have been obtained if the selection had been in the reverse direction. It will be seen that only as regards the last column, which deals with the period when recessives are rare, is there a great difference between different types of selection.

I next considered (Haldane, 1924*b*) the results if the population, instead of being wholly inbred or mating at random, was partly inbred.

If a proportion λ of the population is self-fertilised while the rest mate at random, u_n being the gametic ratio as above, the proportion of recessives is nearly

$$\frac{2 - \lambda + \lambda u_n}{(2 - \lambda)(1 + u_n)^2}$$

and[5]

$$kn = \log_e\left(\frac{u_n}{u_o}\right) + \frac{2}{\lambda}\log_e\left(\frac{2 - \lambda + \lambda u_n}{2 - \lambda + \lambda u_o}\right)$$

so that even when recessives are few, their numbers increase or decrease in a geometrical progression whose common ratio is approximately $1 + (\lambda k)/(2 + \lambda)$.[6] Similarly if a proportion λ of the population is mated to whole brothers or sisters the recessives, when rare, increase or decrease in a geometrical progression whose common ratio is approximately $1 + (\lambda k)/(4 - 3\lambda)$. So a small amount of inbreeding (matings between cousins have a similar but less effect) will enable selection to act on rare recessives.

[5, 6] See Notes by E. G. Leigh Jr. on p. 210 *et seq.*

TABLE VI *Generations required for given Change in Proportion of Dominants*

Gene favoured	Type of selection	Sex	0.001–1%	1–50%	50–99%	99–99.999%
Plasmon	Any	Both	6,921	4,592	4,592	6,921
Autosomal	Ordinary	Both	6,920	4,819	11,664	309,780
Autosomal	Familial	Both	13,841	9,638	23,328	619,560
Autosomal	Gametic	Both	13,831	8,819	6,157	7,112
Sex-linked	Ordinary	Homogametic	6,916	4,668	5,593	10,106
Sex-linked	Ordinary	Heterogametic	6,928	5,164	11,070	20,693
Sex-linked	Homogametic only	Homogametic	10,380	7,228	17,496	464,670
Sex-linked	Homogametic only	Heterogametic	10,392	8,378	153,893	149,860,377
Sex-linked	Heterogametic only	Homogametic	20,746	13,228	9,236	10,668
Sex-linked	Heterogametic only	Heterogametic	20,753	13,785	13,785	20,753

On the other hand assortative mating or selective fertilisation has no appreciable effect. The reason for this is simple. If $u = 999$, there is one recessive in a million in a random mating population, but one dominant in 500 is heterozygous. It does not much matter how the recessives mate, but it is very important that a large proportion of the heterozygous dominants should mate with themselves or one another.

In autopolyploid plants the laws of selection are very similar. With a gene ratio u_n the proportion of recessives in a random mating population is $(1 + u_n)^{-4}$ in a tetraploid, and the rate of change of u_n under slow selection is given by $du_n/dn = k/(1 + u_n)^3$, so that selection is a slower process than in diploid, except when dominants are very few.

EQUILIBRIA INVOLVING ONLY ONE GENE

I have considered two cases in large random-mating populations, one in which the heterozygote is fitter than either homozygote (Haldane, 1926), and one in which the effect of selection is balanced by mutation. If the population, after selection, is in the ratios

$$(1 - K)u_n^2 AA : 2u_n Aa : (1 - k)aa,$$

then

$$\Delta u_n = \frac{u_n(k - Ku_n)}{u_n + 1}$$

so the population is in equilibrium when $u_\infty = k/K$, and stable if k and K are positive. If $k = 1$, i.e. the gene a is lethal, $u_\infty = 1/K$, the population in equilibrium being $1AA : 2KAa$. The equilibrium is fairly quickly approached from both sides. A stable equilibrium is also possible in the case of a sex-linked gene. If K is negative, i.e. dominance incomplete, the homozygote being fitter than the heterozygote, selection is fairly rapid in all stages. A new recessive gene has thus a far greater chance of spreading through the population if it is not completely recessive than if it is so.

A large number of equilibria are possible involving lethal genes which are of advantage when heterozygous. They are of evolutionary importance, because, as pointed out in Chapter II, many chromosomal abnormalities behave like lethal genes. I have nowhere discussed this question fully, because there are very many individual cases to be considered. If the lethal gene operates early enough we have to deal with familial selection. Thus in *Oenothera* many species are heterozygous for genes, groups of genes, or deficiencies which kill off half their pollen grains, half their seeds, or both. The former is of no disadvantage to species which are mainly self-fertilised, the latter of very little. For the numerous seeds from the same plant mostly fall together, and compete with one another, the number surviving in nature being much the same in a species such as *Lamarckiana* where half perish inevitably as the result of lethals, and in *Hookeri* where nearly all survive under very favourable conditions. The net result of my calculations, as of the unpublished calculations of Muller (1930) is that it is hard to see how an *Oenothera*-like condition could arise as the result of selection in an out-breeding organism. The discovery of such a condition in animals would therefore tell against the theory of its evolution by natural selection. So far, however, it has only been found in plants which are usually self-fertilised.

Selection may be, and indeed commonly is, balanced by mutation (Haldane, 1927*b*). Consider a population containing a disadvantageous recessive gene *a*. Then if the probability of A mutating to *a* in each generation is *p*, that of the reverse mutation *q*, instead of $u_{n+1} = [u_n(u_n + 1)]/(u_n + 1 - k)$ we have

$$u_{n+1} = \frac{(1 - p)(u_n^2 + u_n) + q(u_n + 1 - k)}{(1 - q)(u_n + 1 - k) + p(u_n^2 + u_n)}$$

So, if *k*, *p*, and *q* are small equation (3) becomes:

$$\Delta u_n = \frac{ku_n}{u_n + 1} - pu_n(u_n + 1) + q(u_n + 1). \qquad (5)$$

Hence if *k* be positive, at equilibrium $u + 1 = \sqrt{(k/p)}$, and the proportion of recessives is equal to *p/k*. This latter is also the case in a self-fertilising population. In particular if the gene is very disadvantageous, so that *k* is nearly 1, the proportion of abnormals in the population will be roughly equal to the mutation frequency. A similar expression, namely $u = k/3p$ approximately, holds for the proportion of the heterogametic sex carrying a semilethal, *e.g.* males with haemophilia. We can conclude that the frequency of mutation of the corresponding normal gene to that found in haemophilia is of the order of once in a hundred thousand generations, *i.e.* *p* is about 10^{-5} or somewhat more. Similar expressions are obtained for the balance in the case of an unfavourable dominant. If mutation occurs in both directions matters are more complicated, and under certain conditions two different stable equilibria are possible. Equilibrium is always approached fairly rapidly.

RAPID SELECTION

If in equation (3) k is not small we can still solve it. If $k = 1$, we have $u_{n+1} = u_n + 1$, so the proportion of recessives in successive generations is

$$\frac{1}{a^2}, \quad \frac{1}{(a+1)^2}, \quad \frac{1}{(a+2)^2},$$

and so on.

In general we write $v_n = 1/u_n$, so $v_{n+1} - v_n = -kv_n^2/(1 + v_n)$. This is a special case of the equation $\Delta v_n = k\varphi(v_n)$, whose solution is now in the press (Haldane, 1932b). It is (subject to certain conditions)

$$kn = \int_{v_o}^{v_n} w\,dv$$

where

$$w = \sum_{r=1}^{\infty} \frac{k^{r-1}}{r!} f_r(v)$$

and, if we write $y = \varphi(v_n)$, and $y_r = d^r y/dv^r$, then

$$f_1(v) = y^{-1}$$
$$f_2(v) = y^{-1}y_1$$
$$f_3(v) = -\tfrac{1}{2}(y^{-1}y_1^2 + y_2)$$
$$f_4(v) = y^{-1}y_1^3 + 2y_1 y_2$$

and so on.

Hence

$$n = \text{constant} + \frac{u_n}{k} + \frac{\log(1 + 1/u_n)}{1 - k} + \frac{1 - k}{k}\log(1 + u_n)$$

approximately.

This enables us to solve a problem posed by Elton's (1927) work on fluctuations of animal populations. Is intense selection with a coefficient km, but operating only every m generations, more effective than moderate selection of intensity k, but operating in every generation? For moderate values of km this will depend on the rate of change with u_n of the coefficient of k in the above series. Cataclysmic selection is faster when dominants are favoured, and slower otherwise, but the difference is not very great.

SLOW SELECTION INVOLVING SEVERAL GENES

This question has been discussed by Fisher (1930) and by myself (Haldane, 1926) in a more pedestrian but, as I think, sometimes more accurate manner. When a character depends for its existence on the presence of m dominants (*e.g.* colour in *Lathyrus* requires the presence of two), and if $_r u_n$ be the genic ratio for the rth of these genes in the nth generation, then the proportion of dominants

$$y_n = \prod_{r=1}^{m} [1 - (1 + _r u_n)^{-2}],$$

and

$$\frac{dy_n}{dn} = 2ky_n^2 \sum_{r=1}^{m} {_r u_n^{-1}}(2 + _r u_n)^{-2}.$$

On comparing with the single gene case it is seen that selection is always slower.

When (as in allopolyploids) a character is a multiple recessive, the proportion of recessives

$$y_n = \prod_{r=1}^{m} (1 + _r u_n)^{-2},$$

$$\therefore \quad \frac{dy_n}{dn} = -2ky_n^2 \sum_{r=0}^{m} {_r u_n}$$

This is soluble by eliminating s between

$$y_n = \prod_{r=1}^{m} (1 - a_r s)^2$$

and

$$kn = \int \frac{ds}{s y_n}$$

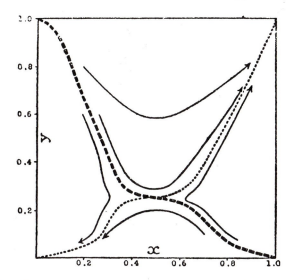

Figure 10. Theoretical effects of selection on a population where the relative fitnesses of four phenotypes are AB, 1; aaB, $1 - 4k$; Abb, $1 - k$; $aabb$, $1 + 11k$. Abscissa, proportion of gene a. Ordinate, proportion of gene b. Trajectories of points representing populations are represented by continuous lines, and boundaries between families of trajectories by dotted lines. (Haldane, 1931b.)

where the a_rs are constants depending on initial conditions. Selection proceeds more slowly than for a single gene.

In the general case where many gene-pairs are present, and every phenotype—perhaps even every genotype—has its own specific fitness, it will ultimately be desirable to represent each population as a point in the analogue of a cube in m-dimensional space. Thus Fig. 10 represents the case of two genes. If the length of the sides of the squares is taken as unity, the point (x, y) represents a random-mating population where two gene pairs A, B, are in the ratios $(1 - x)$A:xa, and $(1 - y)$B:yb. The effect of selection will be to move this point. So a series of points lying on a trajectory will represent the state of the population in succeeding generations. Through each point will pass a single trajectory, and every trajectory will pass to a point of stable equilibrium. Two such groups separated by a boundary are shown in the figure. In three or four dimensions matters are more complicated, and I have so far failed to obtain explicit equations either for the trajectories or their boundaries.

In the two-dimensional case we have two gene-pairs in the ratios

$$u_n A : 1a, \quad v_n B : 1b, \quad \text{and} \quad x = \frac{1}{1 + u_n}, \quad y = \frac{1}{1 + v_n}$$

Then if the relative fitnesses of the four genotypes are:

$$AB \quad 1$$
$$aaB \quad 1 - k_1$$
$$Abb \quad 1 - k_2$$
$$aabb \quad 1 + K$$

$$\therefore \quad \Delta u_n = \frac{u_n[k_1(1 + v_n)^2 - K - k_1 - k_2]}{(1 + u_n)(1 + v_n)^2}$$

approximately, whence $dx/dn = x^2(1 - x)[(K + k_1 + k_2)y^2 - k_1]$, and a similar expression for dy/dn. Putting*

$$a^2 = \frac{k_2}{K + k_1 + k_2}, \quad b^2 = \frac{k_1}{K + k_1 + k_2},$$

we have

$$\frac{dy}{dx} = \frac{y^2(1 - y)(x^2 - a^2)}{(x^2(1 - x)(y^2 - b^2)}$$

---

* The original edition had

$$a^2 = \frac{k_1}{K + k_1 + k_2} \quad \text{and} \quad b_2 = \frac{k_2}{K + k_1 + k_2}.$$

These have been corrected to

$$a^2 = \frac{k_2}{K + k_1 + k_2} \quad \text{and} \quad b^2 = \frac{k_1}{K + k_1 + k_2}.$$

Conversion of b_2 to b^2 corrects a trivial typographic error, but the more substantive exchange of k_1 and k_2 is necessary to derive the expression for dy/dx. The error, which was repeated from Haldane, 1931b, was found by Henry S. Horn while looking for an inconsistency in the interpretation of Figure 10 that had been noticed by James F. Crow. The correction sets the math and the figure to rights, but leaves a need for care in interpretation because the parameter "a" is now associated with the fitness deficit of the Abb genotype alone and the parameter "b" with that of aaB. Haldane's interpretations of the figure, both here and in 1931b, are qualitative, and therefore they are unaffected by the correction.

Footnote to the original edition and correction to the text added by Henry S. Horn, Princeton University 1990.

$$\therefore \quad f(y, b) - f(x, a) = c$$

where

$$f(x, a) = \frac{a^2}{x} - a^2 \log_e x + (a^2 - 1) \log_e (1 - x),$$

and c is determined by the initial conditions. The trajectories given by these equations are divided into four groups by the boundary curves (shown dotted): $f(y, b) - f(x, a) = f(b, b) - f(a, a)$. In the figure $a = \frac{1}{2}, b = \frac{1}{4}$.

MUTATION PRESSURE AS A CAUSE OF EVOLUTION

Fisher (1931) was, I believe, the first to point out the fact that mutation alters the environment in which other genes are placed, and thus the incidence of selection. Any population is riddled with unfavourable genes, both recessive and dominant, each present in a very small proportion of the population, and saved from extinction by mutation. But there are so many of these that a large proportion of an outbred population is at least heterozygous for one of them, as Tschetwerikoff (1927), Jenkin (1928), and others have found experimentally. Now suppose we have two allelomorphic genes A_1 and A_2, which are neutral in presence of the normal gene assortment, i.e. selection does not favour one at the expense of the other. Next suppose that another normal gene B has an allelomorph B' (either dominant, recessive or intermediate) which is a handicap to its possessor. Further, suppose that in the presence of A_2, B' is less harmful than in presence of A_1. Then this will constitute a selective advantage in favour of A_2, and A_2 will tend to replace A_1 in the population.

Fisher (1931) has based a theory of the evolution of dominance on this basis. He believes that abnormal genes are originally intermediate in dominance, rather than recessive. But modifiers are selected which render the heterozygote normal in its viability. I have criticised this theory (Haldane, 1930a) though I believe it to be true in some cases. Fortunately, however, it is susceptible of experimental proof or disproof (Fisher, 1930, p. 62), and since Fisher is undertaking the necessary experiments there is no need to state the arguments for and against his theory here, since at least one of these arguments will be shown to be fallacious in the near future.

Mutation pressure must be a slow cause of evolution, but it certainly cannot be neglected when organisms are in a fairly constant environment over long periods. Among other things it will favour polyploids, and particularly allopolyploids, which possess several pairs of sets of genes, so that one gene may be altered without disadvantage, provided its functions can be performed by a gene in one of the other sets of chromosomes. Occasionally we can point to its probable results in a diploid. Thus in *Primula sinensis* plants possessing the gene *Ch* (for 10 as opposed to 5 calyx teeth, etc.) have fully functional flowers and leaves. But other genes, such as *f* (crimped leaves) and *mp* (maple leaves) produce greater abnormality in *ChCh* and *Chch* than in *chch* plants, thus favouring the gene *ch*.

Selection of a Metrical Character determined by many Genes

Consider an apparently continuously varying character such as human stature. The distribution of such characters is usually normal, *i.e.* according to Gauss' error curve. When a population is in equilibrium it has been shown in several cases that mortality is higher or fertility less in those individuals which diverge most from the mean.

Fisher's (1918) analysis of Pearson's data on the correlation between relatives shows that human stature is inherited as if (apart from rather small environmental influences) it were determined by a large number of nearly completely dominant genes, each acting nearly independently on the character concerned. If there were no dominance the average stature of children would be given by that of their parents. We should get very little further information on it (as we actually do) from a knowledge of the stature of remoter ancestors.

Fisher's (1930) analysis of the effect of selection on such a population involves his theory of the evolution of dominance, which I do not myself hold. His analysis is very greatly simplified if we restrict ourselves, as I shall do here, to the case where all the genes concerned are fully dominant.

Consider a dominant gene A which is present with a genic ratio u_n, *i.e.* the three genotypes are in the proportions $u_n^2 AA : 2u_n Aa : 1aa$.

Let α be the difference between the mean stature of the dominants and recessives. Then the average deviation of the dominants from the general mean of the population must be $a/(u_n + 1)^2$, that of the recessives $-[(u_n^2 + 2u_n)a]/(u_n + 1)^2$.

The dominants will form a normally distributed group with a mean stature exceeding the general mean by $a/(u_n + 1)^2$, where a may, of course, be negative. The standard deviations of the two groups will be equal, but their average divergences from the mean will differ. The group whose mean stature is nearest to that of the general population will be fittest. The two will diverge equally if dominants and recessives are present in equal numbers, *i.e.* $u_n^2 + 2u_n = 1$, or $u_n = \sqrt{2} - 1$. In this case the population is in equilibrium. If u_n exceeds this value there will be more dominants than recessives, and the recessives will, on the whole, be more abnormal and therefore less fit than the dominants. So the proportion of dominants, and hence u_n, will increase. Similarly if u_n is less than $\sqrt{2} - 1$ it will decrease still further as the result of selection. The argument can obviously be extended to populations where complete or partial inbreeding is the rule. Fisher shows that it is also true when dominance is incomplete, in the particular case where the relative unfitness, or coefficient of selection, varies as the square of the mean deviation from the general average.

Hence *a normally distributed population cannot be in stable equilibrium as a result of selection for the characters normally distributed.* This rather sensational fact vitiates a large number of the arguments which are commonly used both for and against eugenics and Darwinism.

If the relative viability or fertility of a population whose mean stature diverges from the mean of the population by $\pm x$ be $1 - cx^2$, which follows from any of a number of simple hypotheses, then

$$k = \frac{ca^2}{(u_n + 1)^4} [(u_n^2 + 2u_n)^2 - 1]$$

$$= \frac{ca^2(u_n^2 + 2u_n - 1)}{(u_n + 1)^2}$$

and

$$\Delta u_n = \frac{ca^2 u_n(u_n^2 + 2u_n - 1)}{(u_n + 1)^3},$$

approximately.[7]

We cannot, however, follow the course of events in such a

[7] See Notes by E. G. Leigh Jr. on p. 210 *et seq.*

population, because the genic ratios for a number of different genes will be varying at once, and hence the mean will vary in an unpredictable manner. In general, however, u_n will increase or decrease until its tendency to do so is checked by mutation in the opposite direction. If p be the probability of A mutating to a in one generation, q that of the reverse process, then

$$\Delta u_n = \frac{u_n(u_n^2 + 2u_n - 1)ca^2}{(u_n + 1)^3} - pu_n(u_n + 1) + q(u_n + 1)$$

For equilibrium this must vanish. So if $x = 1/(1 + u)$ (the proportion of recessive genes), then

$$x^2(1 - x)(1 - 2x^2) + \frac{p(x - 1) + qx}{ca^2} = 0$$

If p and q are small compared with ca^2 this has three roots between 0 and 1, one approximating to $1/\sqrt{2}$ defining an unstable equilibrium, the others near $\sqrt{p/ca^2}$ and $1 - (q/ca^2)$ defining stable equilibrium. Since p and q are small, either dominants or recessives are fairly rare. Hence most of the variance is due to rare and disadvantageous genes whose supply is only kept up by mutation. But only in so far as it includes such genes does a population possess the genetic elasticity which permits it to respond to a change in environment by evolving. It must be remembered that if any gene, apart from its effect on stature, is advantageous in the heterozygous condition, it will tend to an equilibrium with u in the neighbourhood of 1. Probably some at least of the heritable stature differences are due to genes of this class. Not only the well-known vigour of hybrids, but the marked amount of heterozygosis found in selected clones, *e.g.* of fruit trees and potatoes, makes their existence probable.

Fisher next considers what will happen if a population in equilibrium of this type is acted on by selection in favour of, say, a larger size. The numerous rare genes for small size will become still rarer, the rare genes for large size becoming commoner. Such changes, if small, will be reversed when the selection ceases. But some of the rare genes for large size will increase in numbers so much as to pass their former point of unstable equilibrium. They will therefore become very common instead of very rare.

Now if the conditions of selection change back to normal these genes will not return to their original frequency, and the mean stature will have been irreversibly increased. Again, supposing that selection increases the optimum stature of a species by a certain quantity, then when the mean stature reaches the new optimum some genes will be past their point of unstable equilibrium, but still increasing in numbers. The stature will thus, so to speak, overshoot the mark aimed at by selection. We have here, for the first time, an explanation, on strictly Darwinian lines, of useless orthogenesis.

In certain rare cases I have shown (Haldane, 1927b) that this might occur even with regard to a character determined by a single fully dominant gene. But this is only so when selection favours dominants, and three inequalities involving k, p, and q are fulfilled. It may also happen as regards a gene (if such exist) where the heterozygote is less fit than either homozygote. But though these may be subsidiary causes of evolution beyond the optimum, they can have far less importance than the Fisher effect.

Very Rare Characters

So far we have argued as if the populations dealt with were infinite, and, what is more, as if the numbers of both (or all) the competing types were infinite. We will first remove the second restriction, $i.e.$ consider what will happen if there are a few individuals of an abnormal type in a very large population. We will first consider a self-fertilising or apogamous population such as wheat or dandelions.

Let p_r be the probability of an individual leaving r offspring, and let $f(x) = p_0 + p_1 x + p_2 x^2 + \cdots$ Hence $f(1) = 1$, $f(0)$ is the probability of leaving no offspring, and $f'(1)$ is the probable number of offspring. If we start with m individuals, the probability of them leaving r descendants between them is the coefficient of x^r in $[f(x)]^m$. If, after n generations the probability of finding r individuals with a certain character is the coefficient of x^r in F(x), the corresponding probability in the $(n + 1)$th generation is the coefficient of x^r in F[$f(x)$]. Hence after n generations the probability that any given individual will have left r descendants is the coefficient of x^r in $f_n(x)$,[*] $i.e.\ f(f(f(f\ldots f(x)\ldots)))$ the operation being repeated n times, and the probability of extinction is $f_n(0)$. If $f'(1)$ is zero or negative, $i.e.$ if the character is neutral or disadvantageous, then

---

[*] Haldane's original (1932) notation for the iterative function was idiosyncratic and has caused some confusion because it was too readily confused with the integral sign. The conventional modern notation is used here.

$$\operatorname*{Lt}_{n\to\infty} f_n(0) = 1,$$

i.e. the character will ultimately disappear. But Koenigs showed that

$$\operatorname*{Lt}_{n\to\infty} f_n(0)$$

is the root of $x = f(x)$ in the neighbourhood of 0. If $f'(x) = 1 + k$, *i.e.* the character is advantageous, $x = f(x)$ has two and only two real positive roots, one $= 1$, the other lying between 0 and 1, though small if k be small. Hence the new character has always a finite chance of survival (Haldane, 1927*b*). The chance of extinction after n generations $f_n(0)$ may be written l_n. Hence $l_{n+1} = f(l_n)$, an equation solved for a particular $f(x)$ on pp. 115–116.

Fisher (1930) considers the case where $f(x) = e^{(1+k)(x-1)}$, *i.e.* the probabilities form a Poisson series and the probable number of offspring is $1 + k$. This is justifiable in an organism producing a large number of offspring, almost all of which die.

Here the probability of ultimate extinction is given by $x = e^{(1+k)(x-1)}$, or putting $x = 1 - y$, where y is the probability of ultimate survival, $1 - y = e^{-(1+k)y}$

$$\therefore \quad k = \frac{y}{2} + \frac{y^2}{3} + \cdots$$

So if k be small, $y = 2k$, approximately. If in the whole history of a species a new type appears more than $(\log_e 2)/2k$ times, it will probably spread through the species. Exactly the same considerations apply to a rare dominant gene.

Thus suppose a new gene has an advantage measured by $k = 0.001$, and appears by mutation with a frequency 10^{-6}, it must appear $500 \log_e 2$, or 347 times before the odds are in favour of its spreading. This requires the appearance of 347,000,000 individuals. So if we are considering the flea, or even man, the new gene will start off on its conquering career within a single generation. But if we are considering *Elephas indicus*, with a total number of the order of 20,000, and a generation of perhaps forty years on the average, it will be nearly a million years before a new gene is likely to spread to a large enough fraction to be sure of spreading farther. The case of a new recessive gene is much less hopeful. In a sufficiently large population a single recessive mutation has an infinitely small chance of spreading, however favourable it is, simply because selection cannot begin to operate in its favour till two recessive genes are present in the same zygote. But before this happens "blind accident and blundering mischance" will have extinguished the gene. Of course, if there is finite mutation *rate* the number of recessives will increase according to equation (5), k being negative. But mutation, not selection, will take the main responsibility for spreading it until the proportion of recessives is p/k.

FINITE POPULATIONS, RANDOM EXTINCTION

The investigation of the case where the total population is finite has been wholly due to Fisher (1930) and Wright (1931). It presents the most serious difficulties yet met with in this investigation, and indeed some of these have not yet been solved, but the work has already raised some problems of very real mathematical and biological interest.

Let us first consider the suggestion, which is constantly being made, *e.g.* by Hagedoorn (1921) and Elton (1930) that random extinction has been an important cause of evolution. If a population of N individuals possesses variation due to m genes, how much of this will have been lost, we may ask, after n generations? There are 2N sets of chromosomes, so each gene causing variation may be present in any number of these from 1 to $2N - 1$. Let p_r be the probability that a gene is present in r sets of chromosomes (its allelomorph being present in the other $2N - r$). We consider each gene *and* its allelomorph. Let $\varphi(x) = \Sigma\, p_r x^r$, as before, and $\varphi(1)$ is the total number of genes considered, *i.e.* $2m$. Then in the next generation the corresponding function is $\varphi[\,f(x)]$, or in the case of a Poisson series $\varphi(e^{x-1})$. This possesses an absolute term p_0 representing the genes which have vanished, and are therefore present in 0 sets of chromosomes. Let us suppose that the number of gene differences in the population is such that one is lost per generation at each end of the distribution, *i.e.* one gene disappears, and its allelomorph becomes present in all the 2N sets of chromosomes. In a large population this rate of loss will be much the same over very many generations, so the population will be almost in a steady state. Hence the effect of the loss of one gene-difference per generation will be spread through all the values of p_r. If we confine our attention to the values of p_r for which r is not very large, *i.e.* if we take any value of x less than 1, we have

$$\varphi[\,f(x)] - \varphi(x) = 1$$

Now consider a positive number l_n less than 1, and defined by the equations $l_{n+1} = f(l_n)$, $l_0 = 0$

$$\therefore \quad \varphi(l_{n+1}) - \varphi(l_n) = 1$$

This is clearly true if $n = \varphi(l_n)$. Moreover $\varphi(l_n) = \varphi(x)$ is large when x is nearly unity, as we shall see, so we are only concerned with the relation between n and l_n when n is large. Fisher solves the problem when $f(x) = e^{x-1}$, so that $l_{n+1} = e^{l_n - 1}$, $l_\infty = 1$.

Put

$$v_n = \frac{1}{1 - l_n}$$

$$\therefore \quad v_{n+1} = (1 - e^{-1/v_n})^{-1}$$

$$= v_n + \tfrac{1}{2} + \frac{1}{12 v_n} - \frac{1}{720 v_n^2} + \cdots$$

$$\therefore \Delta v_n = \tfrac{1}{2} + \frac{1}{12 v_n} - \frac{1}{720 v_n^2} + \cdots$$

$$\therefore \quad v_n - v_0 = \frac{n}{2} + \frac{1}{12} \sum_{n=0}^{n} \frac{1}{v_n} - \theta$$

where θ does not increase indefinitely with n. But $v_0 = 1$, and after simplification, we find

$$v_n = \frac{n}{2} + \frac{1}{6} \log_e n + c,$$

c being less than unity. When n is large $v_n = n/2$ approximately, so

$$n = 2v_n = \frac{2}{1 - l_n} = \frac{2l_n}{1 - l_n}$$

approximately,

$$\therefore \quad \varphi(n) = \frac{2x}{1 - x} = 2(x + x^2 + x^3 + \cdots)$$

Hence there are approximately one pair of genes in each of the possible frequencies, or $m = 2N$. Fisher gives a much more exact expression for $\varphi(x)$ and a more rigorous proof. If the population is almost entirely self-fertilised, and therefore homozygous, gene-differences disappear twice as fast, and $m = N$. In the first case the fraction of all genes lost per generation is 1/2N, *i.e.* $dm/dn = 1/2N$, and after n generations the number of gene differences is reduced to $e^{-n/2N}$. So a time of $2 \log_e 2N$ or $1.39N$ generations is needed to halve this number. In a numerous species this is a long period even on an astronomical, let alone a geological time scale, in other words random extinction plays no part in evolution. Elton (1930), however, regards it as important in species where an epidemic or other catastrophe periodically reduces the numbers. In such a case we may take for N the minimum number of adults during a cycle, and the length of a cycle, which may vary from 1 to 22 years in the cases considered by him as the "generation." He considers the arctic fox, and takes it that every 3 years the number of this species in Kamschatka is reduced from 800,000 to 80,000. So N = 80,000, and the period needed for an even chance of random extinction of a given gene is 330,000 years. This period takes us back well beyond the last ice-age, *i.e.* to a time when ecological conditions were quite different from those to-day. So random extinction has probably played a very subordinate part in evolution, even in favourable cases.

Other events of the same character, *e.g.* the spread of a new gene from an original single individual to a majority of the species, will require periods of the order of N generations. We cannot say that they have never happened, but we can say that they have played a part quite subordinate compared with that of selection or even mutation.

FINITE POPULATIONS, RANDOM EXTINCTION BALANCED BY MUTATION

Consider, with Fisher (1930), a population as in the last part, save that it is in equilibrium because, in the whole population, one new gene-difference per generation occurs through mutation, and this just balances the loss of variance due to random extinction. An entirely novel mutation is supposed to occur in each generation, so the case is slightly artificial. In general it will only apply to genes which only occur very rarely as the result of mutation, and not to those which mutate frequently. Adopting the terminology of the last section,

$$\varphi[f(x)] - \varphi(x) = 1 - x$$

since the effect of mutation, represented by the right hand side is to diminish the absolute term of $\varphi(x)$ by 1, and to increase p_1, the coefficient of x, by 1. This must just balance the effect of random extinction.

$$\therefore \quad \varphi(l_{n+1}) - \varphi(l_n) = 1 - l_n$$

But, if $l'_n = dl_n/dn$, we have

$$l'_{n+1} = e^{l_n - 1} l'_n$$

$$\therefore \quad l_n - 1 = \log_e l'_{n+1} - \log_e l'_n$$

$$\therefore \quad \varphi(l_{n+1}) + \log_e l'_{n+1} = \varphi(l_n) + \log_e l'_n = c \text{ (a constant)}$$

$$\therefore \quad \varphi(l_n) = c - \log l'(n)$$

But when n is large

$$\frac{1}{l'_n} = \frac{dn}{dl_n} = \frac{2}{(1 - l_n)^2}$$

approximately,

$$\therefore \quad \varphi(l_n) = \text{constant} - 2 \log_e (1 - l_n)$$

$$\therefore \quad \varphi(x) = C + 2 \left(\frac{1}{x} + \frac{1}{2x^2} + \frac{1}{3x^3} + \cdots \right)$$

The total number m of gene-differences $\varphi(1)$, is this series summed to 2N terms, or $2 \log_e (2N) + \text{constant}$. Fisher evaluates the constant as 1.355.

$$\therefore \quad m = 2 \log_e N + 2.741$$

Now if p be the mean mutation frequency, Np new genes are produced per generation, so

$$m = pN(2 \log N + 2.741)$$

i.e. in a large species there should be many more different gene-pairs available than in a small. This refers to neutral genes like those for banding in *Cepea*. But a change of environment may cause them to acquire selective value. Thus a large species, being more variable than a small, will tend to be more plastic under the influence of selection.

FINITE POPULATIONS, ALLOWING FOR RANDOM EXTINCTION, MUTATION, AND SELECTION

Fisher has considered this case, but has only dealt thoroughly with the situation which arises when there is no dominance, and the effect of selection is to make u_n increase or decrease in a geometric series, as always occurs when dominants are rare, or inbreeding intense. I shall do no more than indicate the method of analysis. Defining θ by the equation $\cos \theta = (1 - u)/(1 + u)$, and defining the frequency of θ by $df = y d\theta$, an expression is obtained for $\partial y/\partial n$. This is only valid for intermediate values of θ, and breaks down at $\theta = 0$ or 2π, where the possible values of u are relatively far apart.

The expression for $\partial y/\partial n$ contains terms such as

$$\frac{1}{4N} \frac{\partial^2 y}{\partial \theta^2},$$

expressing the effect of random survival, and a term

$$-\tfrac{1}{2}k \frac{\partial}{\partial \theta} (y \sin \theta)$$

expressing the effect of selection. The relative importance of these terms depends on the value of kN. Only when kN is of the order of 1 or less can the gene be regarded as anything like neutral. Thus probably very few genes are at all neutral at any moment. They are either disadvantageous, and only kept in existence by mutation, or spreading relatively rapidly through the population.

Though Fisher's analysis does not cover the cases of dominance and intense inbreeding, an extension to them will only involve doubling or halving the values of k which render selection definitely effective.

SOCIALLY VALUABLE BUT INDIVIDUALLY DISADVANTAGEOUS CHARACTERS

A study of these traits involves the consideration of small groups. For a character of this type can only spread through the population if the genes determining it are borne by a group of related individuals whose chances of leaving offspring are increased by the presence of these genes in an individual member of the group whose own private viability they lower.

Two simple cases will make this clear. Broodiness is inherited in poultry. In the wild state a broody hen is likely to live a shorter life than a non-broody one, as she is more likely to be caught by a predatory enemy while sitting. But the non-broody hen will not rear a family, so genes determining this character will be eliminated in nature. With regard to maternal instincts of this type selection will presumably strike a balance. While a mother that abandoned her eggs or young in the face of the slightest danger would be ill-represented in posterity, one who, like the average bird, does so under a sufficiently intense stimulus will live to rear another family, which a too devoted parent would not.

In the case of social insects there is no limit to the devotion and self-sacrifice which may be of biological advantage in a neuter. In a beehive the workers and young queens are samples of the same set of genotypes, so any form of behaviour in the former (however suicidal it may be) which is of advantage to the hive will promote the survival of the latter, and thus tend to spread through the species. The only bar to such a spread is the possibility that the genes in question may induce unduly altruistic behaviour in the queens. Genes causing such behaviour would tend to be eliminated.

When we pass to small social groups where every individual is a potential parent, matters are complicated.Consider a tribe or herd of N individuals mating at random, and in the ratios $u_n^2 AA : 2u_n Aa : 1aa$. Let the possession of the recessive character for altruistic behaviour caused by aa decrease the probable progeny of its possessors to $(1 - k)$ times that of the dominants. Let the presence of a fraction x of individuals in the tribe increase the probable progeny of all its members to $(1 + Kx)$ times that of a tribe possessing no recessives, which we may take to be in equilibrium. We will further suppose that a tribe composed entirely of recessives would tend to increase, hence $K > k$.

Now in the next generation the number of the tribe will be increased to

$$N\left[1 + \frac{K}{(1 + u_n)^2}\right].$$

The number x_n of recessive genes will have changed from

$$\frac{N}{1 + u_n} \text{ to } N\left[1 + \frac{K}{(1 + u_n)^2}\right]\frac{(u_n + 1 - k)}{(1 + u_n)^2}$$

approximately. Hence, neglecting kK,

$$\Delta x_n = \frac{N\left(\dfrac{K}{1 + u_n} - k\right)}{(1 + u_n)^2}$$

Hence the number of recessive genes will increase so long as $u_n + 1 > K/k$, *i.e.* recessive genes will only increase as long as they are fairly common. But meanwhile u is increasing, so this process will tend to come to an end. If altruism is dominant, we find that the numbers of the gene for it tend to increase if $(1 + u_n)^2 > K/(K - k)$. In other words the biological advantages of altruistic conduct only outweigh the disadvantages if a substantial proportion of the tribe behave altruistically. If only a small fraction behaves in this manner, it has a very small effect on the viability of the tribe, not sufficient to counterbalance the bad effect on the individuals concerned. If K/k be large, the proportion of altruists need not be great. If $K/k > N$ in the case of a dominant gene, $K/k > \sqrt{N}$ in the case of a recessive, a single altruistic individual will have a net biological advantage. Hence for small values of N selection is at once effective. But in large tribes the initial stages of the evolution of altruism depend not on selection, but on random survival, *i.e.* what in physics is called fluctuation. This is quite possible when N is small, very unlikely when it is large. If any genes are common in mankind which promote conduct biologically disadvantageous to the individual in all types of society, but yet advantageous to society, they must have spread when man was divided into small endogamous groups. As many eugenists have pointed out, selection in large societies operates in the reverse direction.

But the conditions given above, though necessary for the spread of congenital altruism, are far from sufficient. Consider a tribe in which the proportion of altruists is sufficient to cause the number of the gene for it to increase. Even so the other allelomorph will increase still more rapidly. So the proportion of altruists will diminish. The tribe, however, will enlarge, and may be expected ultimately to split, like that of Abram (Genesis xiii. 11). In general this will not mend matters, but sometimes one fraction will get most of the genes for altruism, and its rate of increase be further speeded up. Finally a tribe homozygous for this gene may be produced. These events are enormously more probable if N is small, and endogamy fairly strict. Even when homozygosis is reached, however, the reverse mutation may occur, and is likely to spread. I find it difficult to suppose that many genes for absolute altruism are common in man.

At the risk of repetition I wish to add that the above analysis refers only to conduct which actually diminishes the individual's chance of leaving posterity (such a chance, though small, does exist even for worker bees). A great deal of human conduct which we call altruistic is egoistic from the point of view of natural selection. It is often correlated with well-developed parental behaviour-patterns. Moreover, altruism is commonly rewarded by poverty, and in most modern societies the poor breed quicker than the rich.

ISOLATION

I have considered in some detail (Haldane, 1930*b*) the conditions under which isolation is effective. Suppose a group of a species to be isolated in an environment (say a cave or a desert) where a genotype which in the normal environment is unsuccessful has an advantage measured by k. Further suppose that in each generation a number of migrants of the original type equal to the population of the isolated area multiplied by a fraction l, immigrate into the area considered. What equilibrium, if any, will be reached? I have considered ten cases, but will only describe three in this summary. k and l are throughout supposed to be small.

If the two types do not interbreed, and A is the original type, B the type favoured in the new environment, and u_n their ratio in the nth generation. The proportion of A in the nth generation is $u_n/(u_n + 1)$; in the $(n + 1)$th,

$$\frac{(1 - k)u_n}{1 + u_n} + l$$

$$\therefore \quad \Delta u_n = l(1 + u_n) - ku_n$$

$$\therefore \quad u_\infty = \frac{l}{k - l}$$

i.e. the final ratio is lA:$(l - k)$B, provided $k > l$. The equilibrium is stable.

If zygotes dominant for a single gene are favoured, but recessives immigrate,

$$\Delta u_n = \frac{u_n}{1 + u_n}[k - l(1 + u_n)^2]$$

Hence the final ratio is of $(k - l)$ dominants to l recessives, provided $k > l$. The equilibrium is stable.

If recessives are favoured but dominants immigrate,

$$\Delta u_n = \frac{l(1 + u_n)^2 - ku_n}{1 + u_n}$$

There are two equilibria, a stable where

$$u_\infty = \frac{k - 2l - \sqrt{k(k - 4l)}}{2l}$$

and an unstable where

$$u_\infty = \frac{k - 2l + \sqrt{k(k - 4l)}}{2l}$$

For equilibrium to be possible k must exceed $4l$, and even so, if there are originally too few recessives, u_0 being greater than the unstable value of u_∞, the recessives will be wiped out.

In all cases considered k/l must exceed a certain value before selection can do anything in the face of migration. Even so, where the character favoured in the isolated area is recessive or depends on the co-operation of several dominants, there is an unstable equilibrium, and a sudden rush of immigrants may swamp the isolated population for ever.

WRIGHT'S THEORY

Wright's (1931) very extensive investigation of the problem of evolution was only published after this book was written. It resembles the work of Fisher more than that of Haldane, but, like the latter, considers migration. It is based on Wright's formulae for the decrease of heterozygosis in a population, which in turn depend on the use of path coefficients of correlation. Unfortunately the exposition of this very powerful method would require a good many pages. Wright arrives at formulae analogous and often equivalent to those of Fisher for the distribution of gene ratios in populations under the simultaneous influences of selection, mutation, random survival, and migration. Unfortunately the type of selection considered is almost always one involving no dominance, *i.e.* in which (under the influence of selection alone) the values of u_n in successive generations form a geometrical progression. I suspect that some of his most important theoretical conclusions would no longer hold if dominance were allowed for. This would greatly complicate the mathematical treatment, but I believe that it must be done before full weight can be given to Wright's results.

He concludes that evolution should be slow in populations which are very small, so that 1/N is larger than the average values of p and k (the mutation and selection coefficients). Here he is undoubtedly correct. In such cases there will be little variation on the population on which selection can act. He also holds that evolution would be a slow process in very large populations where 1/N is small compared with p and k. On the other hand medium-sized populations are large enough to be reasonably variable, but not too large to permit of large changes in gene-ratios due to random survival. He holds that this random survival has played a part in evolution much more important than that assigned to it by Fisher or myself. Only a very thorough discussion, which has not yet even begun, can decide which of us is correct.

But Wright's theory certainly supports the view taken in this book that the evolution in large random-mating populations, which is recorded by palaeontology, is not representative of evolution in general, and perhaps gives a false impression of the events occurring in less numerous species. It is a striking fact that none of the extinct species, which, from the abundance of their fossil remains, are well known to us, appear to have been in our own ancestral line. Our ancestors were mostly rather rare creatures. "Blessed are the meek: for they shall inherit the earth."

OTHER INVESTIGATIONS

I have only dealt with such of Fisher's theoretical investigations as seem to me to bear on the questions raised in this book, and have therefore omitted several important topics, *e.g.* the theory of correlations between relatives in a population in equilibrium under selection and mutation. Nor have I mentioned the beautiful work of Volterra (1931) and Lotka (1925) on mathematical ecology, which, however, bears on the struggle between different species rather than between different varieties of the same species.

Much remains to be done, even in the development of the elementary theory which has been followed in my papers. In particular the case of multiple allelomorphism remains for consideration. I do not think that it will lead to any very novel results from the biological point of view, and it is rather involved mathematically. Linkage has generally very little obvious effect, for, as I have shown (Haldane, 1926) a pair of linked genes under selection distribute themselves evenly between the two chromosomes concerned if $l < k$, where $100l$ is their cross-over value and k the coefficient of selection. Fisher (1930, p. 103), however, thinks that natural selection may in certain cases increase or decrease linkage. In organisms where linkage is very intense, *e.g. Orthoptera*, matters are different. Fisher (1931) has begun the investigation of this problem, but its full treatment must await further experimental analysis of the linkage relations concerned.

CONCLUSION

I hope that I have shown that a mathematical analysis of the effects of selection is necessary and valuable. Many statements which are constantly made, *e.g.* "Natural selection cannot account for the origin of a highly complex character," will not bear analysis. The conclusions drawn by common sense on this topic are often very doubtful. Common sense tells us that two bodies attracting one another by gravitation tend inevitably to fall together, which would sometimes be true if the force between them varied as r^{-n}, n exceeding 2. It is not true with the inverse square law. So with selection. Unaided common sense may indicate an equilibrium, but rarely, if ever, tells us whether it is stable. If much of the investigation here summarised has only proved the obvious, the obvious is worth proving when this can be done. And if the relative importance of selection and mutation is obvious, it has certainly not always been recognised as such.

The permeation of biology by mathematics is only beginning, but unless the history of science is an inadequate guide, it will continue, and the investigations here summarised represent the beginning of a new branch of applied mathematics.

Bibliographical References

PAGE

BATESON (1928), "William Bateson, Naturalist," Cambridge . . . 17
de Beer (1930), "Embryology and Evolution," Oxford 80, 81
Berg (1926), "Nomogenesis," London 6, 38, 48
Bernard (1878), "Leçons sur les phenomènes de la vie communes
    aux animaux et aux vegétaux," Paris 20
Betts (1928), "The Iron Age," Oxford 78
Bidder (1930), "Nature," 125, p. 783 64
Birmingham, Bishop of (1930), "Nature," 126, p. 842 7
Blakeslee (1928), "Proc. 5th Int. Cong. Genetics," p. 117
    ("Zeit. Ind. Abst. u. Vererb.," Berlin) 29
Bolk (1926), "Das Problem der Menschenwerden," Jena . . . 15, 68
Brink, (1927), "Genetics," 12, p. 461 67
de Broglie (1930), "An Introduction to the Study of Wave Mechanics,"
    London 85

CALMETTE et Guérin (1924), "Ann. Inst. Past.," 38, p. 371 . . . 51
di Cesnola (1904), "Biometrika," 3, p. 58 48
Chittenden (1928), "Journ. Gen.," 19, p. 285 39
Coleman, (1928), "Victorian Nat.," 44, p. 10 69
Crew (1927), "Organic Inheritance in Man," Edinburgh . . . 63

DARLINGTON (1928), "Journ. Gen.," 19, p. 213 . . . . . 29, 75
    ,,   (1929),   ,,   ,,   20, p. 345 28
    ,,   (1931),   ,,   ,,   24, p. 405 43
    ,,   and Moffett (1930), "Journ. Gen.,", p. 130 . . . 44
Darwin and Wallace (1858), "Journ. Linn. Soc." (July 1) . . . 1
Detlefsen (1914), "Carn. Inst. Wash. Pub.," 205 . . . . 40, 59
    ,,   and Roberts (1918), "Genetics," 3, p. 573 59
Diver (1929), "Nature," 124, p. 183 46

ELTON (1927), "Animal Ecology," London 65, 105
    ,,   (1930), "Ecology and Evolution," Oxford . . . 86, 115, 117
Engledow (1925), "Journ. Agric. Sci.," 15, p. 124 51

FEDERLEY (1913), "Zeit. Ind. Abst. und Vererb.," 9, p. 1 . . . 39
Fisher (1918), "Trans. Roy. Soc. Edin.," 52, p. 399 . . . 26, 110
    ,,   (1930), "The Genetical Theory of Natural Selection,"
        Oxford 6, 92, 94, 106, 109, 110, 114, 115, 117, 125
    ,,   (1931), "Biol. Rev.," 6, p. 345 . . . . . . 109, 125
    ,,   and Ford (1929), "Trans. Ent. Soc. Lond.," 76, p. 378 . . 71
Ford (1930), "Trans. Entom. Soc.," 38, p. 345 57

GAIRDNER (1929), "Journ. Gen.," 21, p. 117 . . . . . 21, 41

Gœthe, "Faust," Pt. II. 3
Goldschmidt (1920), "Mechanismus und Physiologie der
    Geschlechtsbestimmung," Berlin 21
    ,,   (1929), "Biol. Zentralblatt," 49, p. 7 32
Gonsalez (1923), "Am. Nat.," 57, p. 289 54
Goodspeed and Clausen (1922), "Univ. Cal. Pub. Bot.," 11 45
Gregory (1914), "Proc. Roy. Soc.," [B], 87 31

HAGEDOORN (1921), "The relative Value of the Processes
    causing Evolution," Hague 115
Hakansson (1929), "Hereditas," 12, p. 1 28
Haldane (Charlotte) (1936), "Man's World," London 71
    ,,   (1922), "Journ. Gen.," 12, p. 101 42
    ,,   (1924a), "Trans. Camb. Phil. Soc.," 23, p. 19 . . . . 98, 99
    ,,   (1924b), "Proc. Camb. Phil. Soc," (Biol. Sci.), 1, p. 158 . . . 101
    ,,   (1926),   ,,   ,,   ,,   ,,   23, p. 363 . . 77, 103, 106, 125
    ,,   (1927a),   ,,   ,,   ,,   ,,   23, p. 607 93
    ,,   (1927b),   ,,   ,,   ,,   ,,   23, p. 838 . . . 104, 112, 114
    ,,   (1927c), "Biol. Rev.," 2, p. 199 . . . . 36, 39
    ,,   (1927d)," Possible Worlds," London 89
    ,,   (1930a), "Am. Nat.," 64, p. 87 . . . . 73, 109
    ,,   (1930b), "Proc. Camb. Phil. Soc.," 26, p. 220 122
    ,,   (1931a),   ,,   ,,   ,,   ,,   27, p. 131 95
    ,,   (1931b),   ,,   ,,   ,,   ,,   27, p. 137 106–108
    ,,   (1931c), 'Proc. Roy. Inst.," 26, p. 355 46
    ,,   (1932a), "Am. Nat.," 66, p. 5. 68
    ,,   (1932b), "Proc. Camb. Phil. Soc.," (in press) 106
    ,,   (1932c), "Biol. Rev." (in press) 33
Hall (1928), "Journ. Gen.," 20, p. 13 73
Hammarlund (1923), "Hereditas," 4, p. 235 28
Harrison (1920), "Journ. Gen.," 9, p. 195 51
Harrison (1928), "Proc. Roy. Soc.," [B], 102, p. 347 . . . . 32, 83
Heribert-Nilsson (1923), "Hereditas," 4, p. 177 66
Hudson (1919), "A Crystal Age," London 71
Hughes (1931), "Nature," 128, p. 496 32
Huskins (1930), "Genetica," 12, p. 531. 58

JENKIN (1928), "Journ. Gen.," 19, p. 391 109
Johannsen (1909), "Elemente der exakten Erblichkeitslehre,"
    Jena 9, 11, 95
Jollos (1930), "Biol. Zentralblatt," 50, p. 541 32
Jørgensen (1928), "Journ. Gen.," 19, p. 133 30

KIHARA (1929), "Cytologia," 1, p. 1 44
Koller (1930), "Journ. Gen.," 22, p. 103 39
Koltsova (1926), "Journ. Exp. Zool.," 45, p. 301 (and personal
    communication) 74
Kosswig (1929), "Zeit. Ind. Abst. und Vererb.," 52, p. 114 58
Kostoff and Kendall (1929), "Journ. Gen.," 21, p. 113 33

LANCEFIELD (1929), "Zeit. Ind. Abst. und Vererb.," 52, p. 287 . . . 42
Lang (1911),   ,,   ,,   ,,   ,,   ,,   5, p. 97. . . . 46
Lotka (1925), "Elements of Physical Biology," New York 124
Lotsy (1916), "Evolution by means of hybridization," Hague . . . 59

de Vilmorin (1856), "C.R. Ac. Sci.," 1856, 2, p. 1871 8
Volterra (1931), "Leçons sur la théorie mathématique de la
   lutte pour la vie," Paris 124
de Vries (1904), "Species and Varieties: their Origin by
   Mutation," London 17
WADDINGTON (1929), "Journ. Gen.," 21, p. 193 23
Weldon (1895), "Proc. Roy. Soc.," 57, p. 360 48
Wells (1895), "The Time Machine," London 89
,, (1923), "Men like Gods," London 88
Wells, Huxley, and Wells (1931), "The Science of Life," London . . . 78
Wettstein (1924), "Zeit. Ind. Abst. und Vererb.," 33, p. 1 40
,, (1928), "Bibliotheca Genetica," 10, p. 1 . . . . 21, 40
Willis (1922), "Age and Area," Cambridge 4, 16–17
de Winton and Haldane (1932), "Journ. Gen.," (in the press) . . . 21
Wright (1926), "Am. Nat.," 60, p. 552 20
,, (1929), ,, ,, 63, p. 274 73
,, (1931), "Genetics," 16, p. 97 . . . . 53, 115, 123

Afterword by Egbert G. Leigh, Jr.

Here, some topics treated in Haldane's mathematical appendix will be treated more fully. This Afterword is intended to

1. Provide a basis for understanding the mathematical foundations of Haldane's arguments;
2. Relate Haldane's mathematical formalism and techniques to those used more recently, as, for example, by Crow and Kimura (1970) or Crow (1986);
3. Show how Haldane's mathematics are related to more recent developments in evolutionary theory; and
4. Outline further developments of some topics discussed by Haldane.

Where a special case serves to illustrate a more general argument of Haldane's, I will use it.

This afterword will use algebra and elementary calculus. I will introduce calculus in a manner designed to assist the memories of readers who have not seen it for a while, but this Afterword provides no introduction to the calculus.

1 MEASURING THE INTENSITY OF SELECTION

Consider a single locus, with alleles A and a, in a population where, as in human beings, generations overlap and individuals reproduce at a variety of ages. Showing how the intensity of selection is related to differences in mortality and reproductive rates at different ages is still the bugbear of theorists. To simplify matters, let us assume that the allele a is so rare that it occurs essentially only in heterozygotes, and that the number of AA's is unchanging.

Let the probability that a newly born AA individual lives to age x be $L(x)$: if it does so, let the probability that, between ages x and $x + 1$, it will be the parent of a newborn, be $2B(x)$. Here, age is best measured in days for waterfleas and in years for humans and elephants: hereafter, I will refer to humans. As each offspring must have two parents, an individual must have two offspring to replace itself. If M is the maximum age our individuals can attain, then

$$L(O)B(O) + L(1)B(1) + \cdots + L(M)B(M) \tag{1.1}$$

is the average number of offspring per parent of AA individuals, where we credit half the offspring to each parent. We represent this sum by the symbol

$$W_{AA} = \sum_{i=0}^{M} L(i)B(i), \tag{1.2}$$

and refer to W_{AA} as the fitness of AA. If $W_{AA} = 1$, so that the population is just replacing itself, and if the same number of individuals is born each year, then the proportion of individuals in the population of age i or less is $L(i)$. Moreover, if the L's and B's are constant, and if, for some age i, $B(i)$ and $B(i + 1)$ are both nonzero, then the number of births per yer will eventually settle down to a constant rate, as one would expect when individuals reproduce independently, at different ages.

Now let the probability that an Aa newborn lives to age x be $L'(x)$. If it does so, let the probability be $B'(x)$ that, between ages x and $x + 1$, it will in turn be the parent of a newborn which inherits its a gene. Then Aa's fitness is

$$W_{Aa} = \sum_{i=0}^{M} L'(i)B'(i). \tag{1.3}$$

When populations are not constant, the simplest change is by geometric progression. Let us therefore suppose that the total number of Aa's born in year t is R^t. Then $R^{t-i}L'(i)$ individuals living in our population at year t were born i years before, in year $t - i$, and during year t a proportion $B'(i)$ of these i-year-olds become parents of Aa newborns. Thus the total number of Aa's born in year t to individuals of all ages is

$$R^t = \sum_{i=0}^{M} R^{t-i}L'(i)B'(i). \tag{1.4}$$

Dividing both sides of equation (1.4) by R^t, one obtains

$$1 = \sum_{i=0}^{M} R^{-i} L'(i) B'(i). \tag{1.5}$$

If we set $R = 1 - r$, and call r the intensity of selection per year against Aa, then equation (1.5) relates this intensity to the survival and reproductive rates $L'(i)$ and $B'(i)$ of different ages of Aa individuals.

Now suppose that r is very small indeed, so that, during the ages i over which reproduction is concentrated, $R^i = (1 - r)^i$ can be approximated by $1 - ir$. Substituting $1 - ir$ for R^i in equation (1.5), we find

$$1 = \sum_{i=0}^{M} (1 - ir) L'(i) B'(i) \tag{1.6}$$

$$= \sum_{i=0}^{M} L'(i) B'(i) - r \sum_{i=0}^{M} i L'(i) B'(i). \tag{1.7}$$

Using equation (1.3), we find

$$1 - W_{Aa} = -r \sum_{i=0}^{M} i L'(i) B'(i). \tag{1.8}$$

The sum which multiplies r is essentially the average age of reproduction. If we set $K(i) = L(i)B(i)$ and $K(i) - k(i) = L'(i)B'(i)$, replace sums by integrals, and use equations (1.2) and (1.3), we find the formulas on the third page of Haldane's appendix. Thus $1 - W_{Aa}$ measures the intensity of selection against Aa per generation if we define generation as the time elapsed from birth to average age of reproduction.

These formulas assume that the L's and b's do not change from year to year, and that individuals of the same genotype and the same age have equal prospects of reproduction. No analytic theory of the intensity of selection in populations with overlapping generations has yet succeeded in removing these assumptions in a satisfying and useful way.

2 THE RELATIONSHIP BETWEEN THE MAGNITUDE OF A MUTATION AND THE PROBABILITY THAT IT IS ADVANTAGEOUS

Using an argument that Haldane summarizes on pp. 94–95, Fisher (1930a) asserted that the smaller the effect of a mutation, and the simpler the characteristic which it affects, the more likely that mutation is to be favorable. Fisher's argument is very schematic, but it does capture the basis of the prevailing belief that small mutations are more likely than large ones to be favorable. In turn, this belief explains the prevailing assumption that populations evolve by the selection of many small mutations rather than a few "macromutations" of large effect. Invoking the chance that a "macromutation" is favorable is equivalent, for many minds, to invoking a miracle. Leigh (1986, 1987) reviews Fisher's argument and subsequent criticisms of it, and examines rather convincing evidence that a macromutation did play a crucial role in the evolution of cultivated maize (Iltis 1983).

3 HOW DOES THE INTENSITY OF COMPETITION AFFECT THE INTENSITY OF SELECTION?

Consider a population where an individual's survival to maturity is governed by its value u of a certain quantitative characteristic (say, size at birth). Suppose that the value u of this characteristic is governed by a locus with two alleles, A and a, where a is so rare that it occurs almost exclusively in heterozygotes, while the distribution of values of u in the population as a whole is essentially the same as its distribution among AA homozygotes.

Suppose, to be specific, that (a) successive generations of this population are distinct (so that all members of one generation die before any of the next achieve sexual maturity); (b) an individual survives to maturity only if its value of u exceeds the threshold value x; (c) the values of u for AA individuals (and thus, essentially, for the population as a whole) are normally distributed with mean 0 and variance s^2: that is to say, the population of these individuals for which u exceeds the value y, is

$$\frac{1}{s\sqrt{2\pi}} \int_y^\infty du \exp u^2/2s^2 = \frac{1}{\sqrt{2\pi}} \int_{y/s}^\infty dv \exp - \tfrac{1}{2}v^2, \qquad (3.1)$$

where $\exp k = e^k = (2.71828..)^k$ (more will be said about this number e in section 4a, for those who don't know it); (d) among Aa

individuals, u is distributed normally with mean $2L$ and variance $(s + 2M)^2$, that is to say, the proportion of Aa individuals for which u exceeds the value y is

$$\frac{1}{(s + 2M)\sqrt{2\pi}} \int_y^\infty du \exp - \frac{1}{2}\left(\frac{u - 2L}{s + 2M}\right)^2$$

$$= \frac{1}{\sqrt{2\pi}} \int_{\frac{y-2L}{s+2M}}^\infty dv \exp - \tfrac{1}{2}v^2. \quad (3.2)$$

Since $v = (u - 2L)/(s + 2M)$, then, when $u = y$, $v = (y - 2L)/(s + 2M)$. Haldane assumed that the distribution of u had mean $-L$ and variance $(s - M)^2$ for the bearers of one genotype, and mean L and variance $(s + M)^2$ for the bearers of the other. So long as L and M are small compared to s, the two assumptions give nearly identical answers.

Since an individual survives to maturity only if its value of u exceeds the threshold value x, the proportion $1/(z + 1)$ of newborns that survive to maturity is

$$\frac{1}{z + 1} = \frac{1}{\sqrt{2\pi}} \int_{x/s}^\infty dv \exp - \tfrac{1}{2}v^2. \quad (3.3)$$

The ratio z of those who die to those who survive, which Haldane defines as the intensity of competition, is

$$z = \int_{-\infty}^{x/s} dv \exp - \tfrac{1}{2}v^2 \bigg/ \int_{x/s}^\infty dv \exp - \tfrac{1}{2}v^2. \quad (3.4)$$

The intensity of selection, k, is defined in terms of the change in the number of Aa individuals per AA individual from one generation to the next. If this ratio multiplies by $1 + k$ per generation, then k is the intensity of selection, otherwise known as the selective advantage of Aa over AA, or the selective differential between Aa and AA. Notice that if the intensity of selection is k, then Aa individuals bear $1 + k$ times as many mature offspring per parent as do AA, in which case the fitness of Aa relative to AA is said to be $1 + k$.

How does the intensity of competition affect the intensity of selection? If all mature individuals have equal fertility regardless of genotype, then $1 + k$ is simply the ratio of the proportion of Aa, to the proportion of AA, newborns surviving to maturity. Thus

$$1 + k = \frac{1}{\sqrt{2\pi}} \int_{\frac{x-2L}{s+2M}}^{\infty} dv \exp - \tfrac{1}{2}v^2 \Bigg/ \frac{1}{\sqrt{2\pi}} \int_{x/s}^{\infty} dv \exp - \tfrac{1}{2}v^2. \quad (3.5)$$

According to equation (3.3), the denominator is equal to $1/(z + 1)$. Therefore

$$1 + k = \left(\frac{z+1}{\sqrt{2\pi}}\right)\left[\int_{x/s}^{\infty} dv \exp - \tfrac{1}{2}v^2 + \int_{\frac{x-2L}{s+2M}}^{x/s} dv \exp - \tfrac{1}{2}v^2\right]$$

$$= 1 + \frac{z+1}{\sqrt{2\pi}} \int_{\frac{x-2L}{s+2M}}^{x/s} dv \exp - \tfrac{1}{2}v^2. \quad (3.6)$$

Subtracting 1 from both sides of equation (3.6),

$$k = \frac{z+1}{\sqrt{2\pi}} \int_{\frac{x-2L}{s+2M}}^{x/s} dv \exp - \tfrac{1}{2}v^2. \quad (3.7)$$

If L and M are small enough compared to s, then

$$k \approx \frac{z+1}{\sqrt{2\pi}} \left(\frac{x}{s} - \frac{x-2L}{s+2M}\right) \exp - x^2/2s^2$$

$$\approx \frac{z+1}{\sqrt{2\pi}} \frac{2Mx + 2Ls}{s(s+M)} \exp - x^2/2s^2. \quad (3.8)$$

Haldane's counterpart to equation (3.8) (1932, p. 96) lacks the factor $s + M$ (essentially s, for M is negligible compared to s) in the denominator. This is an error.

Suppose now that L is negative, equal to $-J$, while M is positive, so that the rare allele a decreases the average of the quantitative characteristic u, but increases u's variance, among its bearers. Equation (3.2) tells us that the larger x, the greater the intensity z of competition. Since

$$k \approx \frac{z+1}{\sqrt{2\pi}} [(2Mx - 2Js)/s^2] \exp - x^2/2s^2, \quad (3.9)$$

then, if x exceeds Js/M, selection favors Aa over AA. In other words, if competition is sufficiently intense, selection favors the genotype with more variable u, regardless of that genotype's effect on u's average value.

If competition is very intense, so that z is very large, then

$$\frac{1}{z+1} = \frac{1}{\sqrt{2\pi}} \int_{x/s}^{\infty} dv \exp - \tfrac{1}{2}v^2 \approx \frac{s}{x\sqrt{2\pi}} \exp - x^2/2s^2 \quad (3.10)$$

(Feller 1968, equation 1.7 on p. 175). For $x/s = 2.7$ ($z = 285$), this approximation is accurate to within 10%. Equation (3.10) implies that

$$z + 1 \approx \frac{x\sqrt{2\pi}}{s} \exp x^2/2s^2, \quad \ln \frac{z+1}{\sqrt{2\pi}} \approx x^2/2s^2 + \ln x/s. \quad (3.11)$$

If x/s is large enough that $x^2/2s^2$ greatly exceeds $\ln x/s$ ($x^2/2s^2$ is 4.9 times $\ln x/s$ when $x/s = 3.5$ and $z = 4000$), then

$$x/s \approx \sqrt{\frac{2 \ln (z+1)}{\sqrt{2\pi}}} \approx \sqrt{\frac{\ln z^2}{2\pi}}. \quad (3.12)$$

Suppose first that $M = 0$, and that the mutant Aa only differs from AA in the mean of u's distribution among its bearers. Then equation (3.8) becomes

$$k \approx \frac{z+1}{\sqrt{2\pi}} (2Ls/s^2) \exp - x^2/2s^2. \quad (3.13)$$

Equation (3.10) tells us that

$$\exp - x^2/2s^2 \approx \frac{x\sqrt{2\pi}}{s(z+1)};$$

$$k \approx \frac{z+1}{\sqrt{2\pi}} \left(\frac{2L}{s}\right)\left(\frac{x}{s}\right) \frac{\sqrt{2\pi}}{z+1} = \frac{2L}{s} \left(\frac{x}{s}\right). \quad (3.14)$$

Using equation (3.12) to substitute for x/s, we find that

$$k = \frac{2L}{s} \sqrt{(\ln z^2)/2\pi}. \quad (3.15)$$

For large z, selective intensity increases *very* slowly with intensity of competition.

Next, assume that $L = 0$, $M \neq 0$, so that Aa affects the variance, but not the mean, of u's distribution. Then, using equations (3.14) and (3.12), equation (3.8) becomes

$$k \approx \frac{z+1}{\sqrt{2\pi}} (2Mx/s^2) \exp - x^2/2s^2$$

$$\approx \frac{z+1}{\sqrt{2\pi}} (2Mx/s^2) \frac{x\sqrt{2\pi}}{s(z+1)} = 2Mx^2/s^3, \quad (3.16)$$

$$k \approx \frac{2M}{s} \ln (z^2/2\pi). \quad (3.17)$$

When competition is intense, changing the variance of u's distribution affects the intensity of selection rather more than changing its mean.

4 EQUATIONS FOR CHANGES IN GENE RATIO AND GENE FREQUENCY

Consider a locus with alleles A and a in a population where successive generations are distinct.

a. Suppose first that our population is haploid: that is to say, each individual contains only one set of chromosomes (one gene at each locus), so that individuals carry either one A or one a. This could happen either if the population reproduces asexually (uniparentally), so that each individual gets its copy of this gene from its only parent; or if the population reproduces sexually, and each fertilized zygote immediately undergoes "reduction division," ensuring that each individual has equal chances of inheriting its copy of this gene from its mother or from its father.

Let $u(n)$ be the number of A-bearers per a-bearer—the "gene ratio"—in the nth generation of our population. How does this gene ratio change from one generation to the next? If adult A-bearers average W_A A-bearing offspring apiece, while adult a-bearers average W_a a-bearers apiece, then

$$u(n + 1) = (W_A/W_a)u(n) = (W_A/W_a)^n u(1). \qquad (4a.1)$$

Set W_A/W_a equal to $1 + k$, the fitness of A relative to a. Then k is the selective intensity, or the selective advantage of A over a. So we may write

$$u(n + 1) = (1 + k)u(n) = (1 + k)^n u(1). \qquad (4a.2)$$

To learn how many generations are required to shift the gene ratio from an initial value u_I to a final value u_F, set

$$u_F = (1 + k)^n u_I, \quad \ln u_F = \ln u_I + n \ln (1 + k), \qquad (4a.3)$$

where ln denotes "natural logarithm", logarithm to the base $e = 2.71828 \dots$. We will soon be explaining the significance of this very curious number. The number n of generations required to shift the gene ratio from u_I to u_F is

$$n = [\ln u_F - \ln u_I]/\ln (1 + k) = \ln (u_F/u_I)/\ln (1 + k). \qquad (4a.4)$$

Equation (4a.2) implies that

$$u(n + 1) - u(n) = ku(n). \qquad (4a.5)$$

Out of that kind of whimsy, which some nonmathematicians find so vilely annoying, let us suppose that $u(n)$ is a continuous function of the real number n, rather than an expression that makes sense only when n is a whole number. If k is very small compared to 1, we may write

$$u(n + 1) - u(n) = \frac{u(n + 1) - u(n)}{(n + 1) - n}$$

$$\approx \lim_{h \to 0} \frac{u(n + h) - u(n)}{h} = \frac{du}{dn} \approx ku, \qquad (4a.6)$$

$$u(n) \approx u(0) \exp(kn) = u(0) e^{kn}. \qquad (4a.7)$$

It is now time to say something about this number $e = 2.7182818\ldots$, which is at once the base of the "natural" or "Napierian" logarithms, and involved in the solution to the "differential equation" $du/dn = ku$. This function e^x is an "exponential function" $E(x)$ that maps addition into multiplication: that is, such functions $E(x)$ have the property that $E(x + y) = E(x)E(y)$ for all numbers x and y. Since $E(x + 0) = E(x)E(0) = E(x)$, $E(0) = 1$. If, for sufficiently small h, $E(h)$ is very nearly $1 + h$, then $E(h) = e(h) = (2.71828\ldots)^h$. The purist would say that if, for sufficiently small h,

$$\lim_{h \to 0} \frac{E(h) - 1}{h} = 1, \qquad (4a.7a)$$

then

$$E(x) = e^x = \lim_{h \to 0} (1 + xh)^{1/h}$$

$$= 1 + x + \frac{x^2}{2!} + \frac{x^3}{3!} + \ldots + \frac{x^n}{n!} + \ldots, \qquad (4a.7b)$$

where $n! = n(n - 1)(n - 2) \ldots (3)(2)(1)$.
Similarly, if $du/dn = ku$, then, for sufficiently small h,

$$\frac{u(h) - u(0)}{h} \approx ku(0);$$

$$u(h) \approx u(0)(1 + hk);$$

$$u(n) \approx u(0)(1 + hk)^{n/h} \tag{4a.7c}$$

$$u(n) = u(0) \lim_{h \to 0} (1 + hk)^{n/h} = u(0)e^{kn}. \tag{4a.7d}$$

Finally, since the natural logarithm is the inverse of the function $e(x) = e^x$,

$$\ln e^x = x. \tag{4a.7e}$$

Logarithms "convert multiplication into addition," which is why, in the days before pocket calculators, they were so useful.

Shifts between difference equations like $u(n + 1) - u(n) = ku(n)$ and differential equations like $du/dn = ku$ are standard practice when k is small. Haldane frequently makes such shifts. To evaluate the error involved, let us first compare equation (4a.2) with equation (4a.7). The predicted values of $u(n)$ differ by over 10% when $k = 0.05$ and $n = 100$, but the approximation is better for smaller values of k (table 4.1).

Now let us consider the time required for selection to shift the gene ratio from u_I to u_F. According to equation (4a.7),

$$u_F = u_I \exp(kn), \quad \ln u_F = kn + \ln u_I, \tag{4a.8}$$

$$n = (1/k) \ln (u_F/u_I). \tag{4a.9}$$

TABLE 4.1 *Difference and Differential Equations Compared*

A. Value of $u(n)$, assuming $u(0) = 1$, for various values of k and n

	$k = 0.001$	0.01	0.05	0.1	0.5
	$n = 5000$	500	100	50	10
$u(n) = u(0)(1 + k)^n$	148	148	148	148	148
$u(n) = u(0)e^{kn}$	148	145	132	117	58

B. Number n of generations required to shift the gene ratio from u_I to $5u_I$, for various values of k

	$k = 0.001$	0.01	0.05	0.1	0.5
$n = \ln (u_F/u_I)/\ln (1 + k)$	5002	502	102	52	12
$n = (1/k) \ln (u_F/u_I)$	5000	500	100	50	10

As Haldane remarks, the time required to shift the gene ratio from u_I to u_F is inversely proportional to the intensity of selection. The differential equation gives much more accurate answers for the time required to shift the gene ratio by a given amount than for the gene ratio's value at a given time (table 4.1). Exponentiation amplifies errors; logarithms cover them.

The frequency in our population of the allele A is the proportion q of its genes at this locus that are A. If there are uA genes per a gene in the population, then A's frequency is

$$q = u/(1 + u), \qquad (4a.10)$$

while a's frequency is $1 - q = 1/(1 + u)$. Nowadays, gene frequencies are more usually employed than gene ratios, so I will show how to pass from one approach to the other.

When the fitness of A relative to a is $1 + k$, there are two ways to calculate $q(n + 1) - q(n)$. The first is to remark that, since $u(n + 1) = u(n)(1 + k)$,

$$q(n + 1) = \frac{u(n)(1 + k)}{1 + u(n)(1 + k)}$$

$$= \frac{u(n)(1 + k)}{[1 + u(n)]\{1 + ku(n)/[1 + ku(n)]\}}. \qquad (4a.11)$$

Setting $u(n)/[1 + u(n)] = q(n)$, we may rewrite equation (4a.11) as

$$q(n + 1) = \frac{q(n)(1 + k)}{1 + kq(n)}, \qquad (4a.12)$$

$$q(n + 1) - q(n) = \frac{q(n)(1 + k)}{1 + kq(n)} - \frac{q(n)[1 + kq(n)]}{1 + kq(n)}$$

$$= \frac{kq(n)[1 - q(n)]}{1 + kq(n)}. \qquad (4a.13)$$

If k is small, equation (4a.13) may be approximated by the differential equation, so familiar in population genetics,

$$\frac{dq}{dn} = kq(1 - q). \qquad (4a.14)$$

The second approach is to say that, if A-bearers average W_A mature offspring per parent, and a-bearers W_a, then

$$q(n + 1) = W_A q(n)/\bar{W}(n),$$

$$q(n + 1) - q(n) = [W_A - \bar{W}(n)]q(n)/\bar{W}(n), \qquad (4a.15)$$

where the mean fitness $\bar{W}(n)$ of the population at generation n is

$$\bar{W}(n) = W_A q(n) + W_a[1 - q(n)]. \qquad (4a.16)$$

Dividing numerator and denominator of equation (4a.15) by W_a, and setting $W_A/W_a = 1 + k$, we find

$$q(n + 1) = \frac{q(n)(1 + k)}{q(n)(1 + k) + 1 - q(n)} = \frac{q(n)(1 + k)}{1 + kq(n)}, \qquad (4a.17)$$

which is just equation (4a.12).

Next, let us calculate $\bar{W}(n + 1) - \bar{W}(n)$, the change per generation in the population's mean fitness. Using equation (4a.16) for $\bar{W}(n)$ and equation (4a.15) for $q(n + 1) - q(n)$, we find that

$$[\bar{W}(n + 1) - \bar{W}(n)]/\bar{W}(n) = [q(n + 1) - q(n)](W_A - W_a)/\bar{W}(n)$$

$$= q(n)[1 - q(n)](W_A - W_a)^2/\bar{W}^2(n)$$

$$= q(n)[1 - q(n)]k^2/[1 + kq(n)]^2. \quad (4a.18)$$

The genic variance $V_g/\bar{W}^2$ in $W/\bar{W}$, the mean square deviation of relative genic fitnesses from the population average, is

$$(W_A - \bar{W})^2 q/\bar{W}^2 + (W_a - \bar{W})^2(1 - q)/\bar{W}^2$$

$$= [(W_A - W_a)^2(1 - q)^2 q + (W_A - W_a)^2 q^2(1 - q)]/\bar{W}^2$$

$$= q(1 - q)(W_A - W_a)^2/\bar{W}^2. \qquad (4a.19)$$

Therefore we may set

$$[\bar{W}(n + 1) - \bar{W}(n)]/\bar{W}(n) = V_g(n)/\bar{W}^2(n). \qquad (4a.20)$$

We may also derive equation (4a.20), using equation (4a.15) and its counterpart for $1 - q(n + 1)$ to set

$$\bar{W}(n + 1) = W_A q(n + 1) + W_a[1 - q(n + 1)]$$

$$= W_A^2 q(n)/\bar{W}(n) + W_a^2[1 - q(n)]/\bar{W}(n), \qquad (4a.20a)$$

$$[\bar{W}(n + 1) - \bar{W}(n)]/\bar{W}(n)$$

$$= \{W_A^2 q(n) + W_a^2[1 - q(n)] - \bar{W}^2(n)\}/\bar{W}^2(n)$$

$$= V_g(n)/\bar{W}^2(n). \qquad (4a.20b)$$

Equation (4a.20) is a prototype of Fisher's (1930) "fundamental theorem of natural selection." Fisher emphasized this theorem because it spotlighted heritable variance in relative fitness as the quantity that governed evolutionary progress. Haldane ignored the theorem. Lewontin (1970), no friend of evolutionary progress, deprecated the theorem, and Turner (1970) ridiculed it. This theorem appears here because it will greatly simplify later calculations. Leigh (1987) discusses a view of what Fisher wished his theorem to mean; Price (1972) and Ewens (1989) assess more precisely what Fisher's version of the theorem actually meant; and Barton and Turelli (1987) and Turelli and Barton (1990) discuss more general versions of the relation between the change in fitness (or any other quantitative characteristic) and the heritable variation therein.

Finally, let us write equation (4a.13) in a new way, not used by Haldane, but useful for explaining propositions of his. Let $\partial/\partial q$ denote the "partial derivative with respect to q", a derivative taken holding all other variables constant. Then

$$[\partial \bar{W}/\partial q]/\bar{W} = \partial \ln \bar{W}/\partial q = (W_A - W_a)/\bar{W} = k/(1 + kq). \qquad (4a.21)$$

We may accordingly express equation (4a.13) for $q(n + 1) - q(n)$, following Wright (1937),

$$q(n + 1) - q(n) = q(n)[1 - q(n)]k/[1 + kq(n)]$$
$$= q(n)[1 - q(n)]\, \partial \ln \bar{W}/\partial q. \qquad (4a.22)$$

Equation (4a.22) and its analogues are of very great generality (Barton and Turelli 1987). For diploids and m-ploids respectively,

$$q(n + 1) - q(n) = \tfrac{1}{2}q(n)[1 - q(n)]\, \partial \ln \bar{W}/\partial q, \qquad (4a.23)$$

$$q(n + 1) - q(n) = (1/m)q(n)[1 - q(n)]\, \partial \ln \bar{W}/\partial q. \qquad (4a.24)$$

In a beautiful paper, Lande (1976a) discusses a generalization of equation (4a.23).

b. Now consider a diploid population, where AA individuals average W_{AA} offspring destined to mature, apiece, while Aa and aa average W_{Aa} and W_{aa} apiece. Let the frequencies of AA, Aa, and aa genotypes among the adults of generation n be $P(n)$, $2Q(n)$, and $R(n)$, respectively. Then A's frequency $q(n)$ is $P(n) + Q(n)$, a's frequency $1 - q(n)$ is $Q(n) + R(n)$, and the gene ratio, the number of A's per a, is

$$u(n) = q(n)/[1 - q(n)] = [P(n) + Q(n)]/[Q(n) + R(n)]. \qquad (4b.1)$$

Suppose now that these adults mate at random, regardless of genotype, and that all matings are equally fertile. Then the frequencies of AA, Aa, and aa zygotes formed by the individuals of generation n are, according to the "Hardy-Weinberg Law," $q^2(n)$, $2q(n)[1 - q(n)]$, and $[1 - q(n)]^2$, respectively (table 4.2). Then

TABLE 4.2 *The "Hardy–Weinberg Law"*

A. Random Mating Among the Genotypes

		Frequency of genotype among males		
Frequency of genotype among females (below)		AA P	Aa $2Q$	aa R
AA	P	$P^2 AA$	$PQ\ AA$ $PQ\ Aa$	$PR\ Aa$
Aa	$2Q$	$PQ\ AA$ $PQ\ Aa$	$Q^2 AA$ $2Q^2 Aa$ $Q^2 aa$	$QR\ Aa$ $QR\ aa$
aa	R	$PR\ Aa$	$QR\ Aa$ $QR\ aa$	$R^2 aa$

NOTES: At the intersection of each row and column are the frequencies of zygotes of the genotypes indicated that arise from matings between fathers of the "column" genotype and mothers of the "row" genotype, assuming that mating is random and that $AA \times AA$ produce only AA; $AA \times Aa$ produce $\frac{1}{2}AA$ and $\frac{1}{2}Aa$; $AA \times aa$ produce only Aa; $Aa \times Aa$ produce $\frac{1}{4}AA$, $\frac{1}{2}AA$, and $\frac{1}{4}aa$; and so forth.

Total frequencies of zygotes of each genotype are

AA: $P^2 + PQ + PQ + Q^2 = (P + Q)^2 = q^2$
Aa: $PQ + PR + PQ + 2Q^2 + QR + PR + QR = 2(P + Q)(Q + R) = 2q(1 - q)$
aa: $Q^2 + QR + QR + R^2 = (Q + R)^2 = (1 - q)^2$.

B. Random Mating among the Gametes

Frequency of		A	a	among eggs
A		q	$1 - q$	
a	q	$q^2 AA$	$q(1 - q)\ AA$	
among sperm	$1 - q$	$q(1 - q)\ AA$	$(1 - q)^2 aa$	

NOTES: Total frequencies of AA, Aa, and aa zygotes are q^2, $2q(1 - q)$, and $(1 - q)^2$, respectively. Note that the zygotic frequencies calculated in part A assuming random mating of genotypes are the same as those calculated in part B assuming random mating of gametes.

$$u(n + 1) = \frac{q(n)[1 - q(n)]W_{Aa} + q^2(n)W_{AA}}{q(n)[1 - q(n)]W_{Aa} + [1 - q(n)]^2 W_{aa}}$$

$$= \frac{q(n)}{1 - q(n)} \left[\frac{q(n)W_{AA} + [1 - q(n)]W_{Aa}}{q(n)W_{Aa} + [1 - q(n)]W_{aa}} \right]. \quad (4b.2)$$

Dividing numerator and denominator by $1 - q(n)$, and setting $u(n) = q(n)/[1 - q(n)]$, we find

$$u(n + 1) = u(n)[W_{AA}u(n) + W_{Aa}]/[W_{Aa}u(n) + W_{aa}]. \quad (4b.3)$$

For our first example, suppose that $W_{aa} = 0$, and that $W_{AA} = W_{Aa}$. Then a is said to be a recessive lethal. In this case,

$$u(n + 1) = u(n)[1 + u(n)]/u(n)$$

$$= 1 + u(n) = n + u(1), \quad (4b.4)$$

$$1 + k = [1 + u(n)]/u(n) = 1/q(n);$$

$$k = 1/u(n) = [1 - q(n)]/q(n). \quad (4b.5)$$

Here, k is the intensity of selection, which in this example depends on the gene ratio. Selection against a recessive lethal changes the gene ratio arithmetically, rather than geometrically as in equation (4a.2). The smaller the frequency $1 - q(n)$ of the recessive allele, the greater the proportion of the alleles that are protected by their association with A in heterozygotes, and the weaker the selective intensity k.

Now let us return to equation (4b.2) and set $W_{AA} = (1 + K)W_{aa}$, $W_{Aa} = (1 + hk)W_{aa}$. Then

$$u(n + 1) = u(n)\frac{(1 + K)u(n) + 1 + hK}{(1 + hK)u(n) + 1}. \quad (4b.6)$$

If $1 + K = (1 + hK)^2$, then

$$u(n + 1) = u(n)(1 + hK)$$

$$= u(n)(1 + K)^{1/2} = u(1)(1 + K)^{n/2}. \quad (4b.7)$$

Here, as in equation (4a.2), $u(n)$ changes by geometric progression. According to equation (4a.4), the number of n of generations required for the gene ratio to change from an initial ratio u_I to a final ratio u_F is

$$n = \ln (u_I/u_F)/\ln (1 + hK) = 2 \ln (u_I/u_F)/\ln (1 + K). \quad (4b.8)$$

According to equation (4b.4), the time required for selection against a recessive lethal to shift the gene ratio from u_I to u_F is given by

$$u_F = u_I + n; \quad n = u_F - u_I. \tag{4b.9}$$

If u_F is large enough, $u_F - u_I$ exceeds $2 \ln (u_F/u_I)/\ln (1 + K)$, which shows once again how slowly selection against a recessive lethal decreases its frequency once it is already rare.

How does inbreeding affect the intensity of selection against a recessive lethal? Suppose, to be specific, that a fraction f of each genotype are self-fertilized (selfed), while the remainder mate at random. Let the frequencies of AA and Aa among the adults of generation n be P and $2Q$, respectively ($R = 0$ since no aa survive to adulthood, and $P + 2Q = 1$). Then $q = P + Q$, $1 - q = Q$. The frequencies of AA, Aa, and aa zygotes arising from random mating are $(1 - f)q^2$, $(1 - f)2q(1 - q)$, and $(1 - f)(1 - q)^2$, respectively (table 4.2). Since selfed AA produce only AA, while selfed Aa produce $\frac{1}{4}AA$, $\frac{1}{4}AA$, and $\frac{1}{4}aa$, the frequencies of AA, Aa, and aa zygotes contributed by selfing are $f(P + \frac{1}{2}Q)$, fQ, and $\frac{1}{2}fQ$, respectively. When Aa is rare, randomly mating Aa produce almost no aa offspring, because nearly all these Aa mate with AA, but a quarter of the offspring of selfed Aa are aa, regardless of how rare a may be.

If many individuals are heterozygous for a harmful recessive allele at one of their loci (and, considering how difficult it is to expel a harmful recessive from a population, who is to say that this is unlikely?), there is a real advantage to outbreeding, to avoiding matings with relatives who might carry the same allele. Returning to the effect of inbreeding on the intensity of selection against a harmful recessive, the gene ratio $u(n + 1)$ among the adults of generation $n + 1$ will be.

$$u(n + 1) = \frac{(1 - f)[q^2 + q(1 - q)] + f(P + \frac{1}{2}Q) + \frac{1}{2}fQ}{(1 - f)q(1 - q) + \frac{1}{2}fQ}. \tag{4b.10}$$

Since $q^2 + q(1 - q) = q = P + Q$, and $Q = 1 - q$,

$$u(n + 1) = \frac{q}{(1 - f)q(1 - q) + \frac{1}{2}f(1 - q)}$$

$$= \frac{q}{1 - q}\left[\frac{1}{(1 - f)q + \frac{1}{2}f}\right]. \tag{4b.11}$$

Remembering that $q/(1 - q) = u(n)$, $q = u(n)/[1 + u(n)]$, and setting $1/(1 + k)$ equal to $(1 - f)q + \frac{1}{2}f$, where k is the intensity of selection, we find that

$$u(n + 1) = \frac{u(n)}{(1 - f)q + \frac{1}{2}f} = u(n)\left[1 + \frac{(1 - f)(1 - q) + \frac{1}{2}f}{(1 - f)q + \frac{1}{2}f}\right]$$

$$= u(n)\left[1 + \frac{(1 - f) + \frac{1}{2}[1 + u(n)]f}{(1 - f)u(n) + \frac{1}{2}[1 + u(n)]f}\right]. \qquad (4b.12)$$

When $u(n)$ is much larger than 1, $k = f/(2 - f)$.

If a fraction f of the individuals in each generation are self-fertilized while the remainder mate at random, and if AA, Aa, and aa are all equally fit, then the frequencies P, $2Q$, and R of AA, Aa, and aa change from generation to generation as follows:

$$P(t + 1) = f[P(t) + \tfrac{1}{2}Q(t)] + (1 - f)q^2(t),$$

$$Q(t + 1) = \tfrac{1}{2}fQ(t) + (1 - f)q(t)[1 - q(t)],$$

$$R(t + 1) = f[R(t) + \tfrac{1}{2}Q(t)] + (1 - f)[1 - q(t)]^2. \quad (4b.12a)$$

The frequency $q(t + 1)$ of A in generation $t + 1$ is

$$P(t + 1) + Q(t + 1) = f[P(t) + Q(t)] + (1 - f)q(t) = q(t). \quad (4b.12b)$$

In the absence of selection, inbreeding does not change allele frequencies: q is constant. Setting $P + \frac{1}{2}Q = \frac{1}{2}(P + q)$, we find that when genotypic frequencies attain equilibrium,

$$P = \tfrac{1}{2}f(P + q) + (1 - f)q^2 = [2(1 - f)q^2 + fq]/(2 - f),$$

$$Q = \tfrac{1}{2}fQ + (1 - f)q(1 - q) = 2(1 - f)q(1 - q)/(2 - f),$$

$$R = \tfrac{1}{2}f(R + 1 - q) + (1 - f)(1 - q)^2$$

$$= [2(1 - f)(1 - q)^2 + f(1 - q)](2 - f). \qquad (4b.12c)$$

Now let the fitnesses of AA, Aa, and aa be 1, 1, and $1 - K$, where K is so small that P, Q, and R bear the same relation to q that they do when $K = 0$. Then

$$1 - q(t + 1) = \frac{1 - q(t) - KR(t)}{1 - KR(t)}. \qquad (4b.12d)$$

If $q(t + 1)$ is very close to 1, then $[1 - q(t)]^2$ is negligible compared to $1 - q(t)$, $R(t)$ is nearly $f[1 - q(t)]/(2 - f)$, and

$$1 - q(t + 1) \approx [1 - q(t)]\left[1 - \frac{Kf}{2 - f}\right], \qquad (4b.12e)$$

not $[1 - q(t)][1 - kf/(2 + f)]$ as given in Haldane (1924, p. 159) and Haldane (1932, p. 101). When a fraction $1 - f$ of each generation mate at random, while a fraction f mate with full siblings, the frequency of a rare recessive allele, whose homozygotes suffer a slight selective disadvantage K, declines by the factor $1 - f/(4 - 3f)$ per generation. To show this, one starts by relating the frequencies of the different types of matings ($AA \times AA$, $AA \times Aa$, etc.) to A's frequency, in the absence of selection.

These results may be generalized in terms of Wright's coefficient of inbreeding (Wright 1932; Crow and Kimura 1970). At our locus, the coefficient of inbreeding is said to be F if a gamete has probability F of uniting with another gamete carrying the same allele, and probability $1 - F$ of mating at random. Thus

$$P(t + 1) = Fq(t) + (1 - F)q^2(t),$$

$$Q(t + 1) = (1 - F)q(t)[1 - q(t)],$$

$$R(t + 1) + F[1 - q(t)] + (1 - F)[1 - q(t)].^2 \qquad (4b.12f)$$

If a fraction f of each generation are selfed, while the remainder mate at random, then, identifying the expressions for Q in equations (4b.12c) and (4b.12f), we find that $F = f/(2 - f)$.

Let us now suppose that the relative fitnesses of AA, Aa, and aA are $1 + K$, $1 + hK$, and 1, respectively and that individuals mate at random. If $q(n + 1)$ is the frequency of A in generation $n + 1$, we may write, in analogy with equation (4a.12),

$$q(n + 1) = \frac{q^2(n)(1 + K) + q(n)[1 - q(n)](1 + hK)}{q^2(n)(1 + K) + 2q(n)[1 - q(n)](1 + hK) + [1 - q(n)]^2}$$

$$= \frac{q(n) + Kq^2(n) + hKq(n)[1 - q(n)]}{1 + Kq^2(n) + 2hKq(n)[1 - q(n)]}, \qquad (4b.13)$$

$$q(n + 1) - q(n)$$

$$= q(n)[1 - q(n)]\left\{\frac{Kq(n) + hK[1 - 2q(n)]}{1 + Kq^2(n) + 2hKq(n)[1 - q(n)]}\right\}. \qquad (4b.14)$$

Notice that

$$\bar{W}/W_{aa} = q^2(n)(1 + K) + 2q(n)[1 - q(n)](1 + hK) + [1 - q(n)]^2. \tag{4b.15}$$

Since $\partial \ln \bar{W}/\partial q = \partial \ln (\bar{W}/W_{aa})/\partial q$, we find, with Wright (1937), that

$$\partial \ln \bar{W}/\partial q = 2\{Kq(n) + hK[1 - 2q(n)]\}/\bar{W}, \tag{4b.16}$$

$$q(n + 1) - q(n) = \tfrac{1}{2}q(n)[1 - q(n)] \partial \ln \bar{W}/\partial q. \tag{4b.17}$$

Finally, let us calculate $[\bar{W}(n + 1) - \bar{W}(n)]/\bar{W}(n)$. Equation (4b.13) may be rewritten as

$$q(n + 1) = W_A q(n)/\bar{W}(n), \ 1 - q(n + 1) = W_a[1 - q(n)]/\bar{W}(n), \tag{4b.18}$$

where

$$W_A(n) = 1 + Kq(n) + hK[1 - q(n)],$$
$$W_a(n) = 1 + hKq(n), \tag{4b.19}$$

$$\bar{W}(n) = W_A(n)q(n) + W_a(n)[1 - q(n)]. \tag{4b.20}$$

The genic variance V_g in fitness at this locus is

$$
\begin{aligned}
V_g &= q(W_A - \bar{W})^2 + (1 + q)(W_a - \bar{W})^2 \\
&= q[(1 - q)(W_A - W_a)]^2 + (1 - q)[q(W_A - W_a)]^2 \\
&= q(1 - q)(W_A - W_a)^2 \\
&= q(1 - q)[K(1 - h)q + hK(1 - q)]^2. \tag{4b.21}
\end{aligned}
$$

The change in relative mean fitness from generation n to generation $n + 1$ is

$$
\begin{aligned}
[\bar{W}(n + 1) &- \bar{W}(n)]/\bar{W}(n) \\
&= [q(n + 1) - q(n)][W_A(n) - W_a(n)]/\bar{W}(n) \\
&= q(n)[1 - q(n)][W_A(n) - W_a(n)]^2/\bar{W}^2(n) \tag{4b.22}
\end{aligned}
$$

$$[\bar{W}(n + 1) - \bar{W}(n)]/\bar{W}(n) = V_g(n)/\bar{W}^2(n). \tag{4b.23}$$

Equation (4b.23) suggests that the genic variance in relative fitness governs the population's response to natural selection. This raises the issue of what maintains genetic variation in natural populations, the topic of the next section.

5 MAINTENANCE OF GENETIC VARIATION

In a uniform environment, where the relative fitnesses of different genotypes are unchanging, there are three basic processes by which populations maintain genetic variation: heterosis, the circumstance where heterozygotes are more fit than homozygotes of either genotype; the balance between the recurrence of harmful mutations and their elimination by natural selection; and the balance between the origin of "neutral" mutations with no effect on fitness, and their eventual disappearance through random changes in frequency. The relative importance of these three processes has been the subject of rather acrimonious debate (Lewontin 1974; Kimura 1983a). In addition, environmental heterogeneity may maintain genetic diversity if the order of fitnesses of different genotypes differs in different environments (that is to say, if there is genotype-environment interaction: Felsenstein 1976; Gillespie and Turelli 1989).

First, let us consider heterosis. Imagine a single locus, with alleles A and a, in a diploid population where successive generations are distinct. Let the relative fitnesses of AA, Aa, and aa be $1 - K$, 1, and $1 - u^*K$, respectively, where K is positive. Using equation (4b.3), let us write

$$u(n + 1) = u(n) \left[\frac{(1 - K)u(n) + 1}{u(n) + 1 - u^*K} \right]. \tag{5.1}$$

Set

$$\frac{(1 - K)u(n) + 1}{u(n) + 1 - u^*K} = 1 + k; \quad k = \frac{K[u^* - u(n)]}{u(n) + 1 - u^*K}. \tag{5.2}$$

Here, the selective intensity k depends on the gene ratio u. The gene ratio ceases to change, that is to say, $u(n + 1) = u(n)$, when $k = 0$. Equation (5.2) shows that $k = 0$ when $u(n) = u^*$. Moreover, when $u(n)$ exceeds u^*, k is negative, and u decreases, while if u^* exceeds $u(n)$, k is positive, and u increases. Thus the gene ratio tends toward the equilibrium u^*: heterosis maintains both alleles in the population in "stable polymorphism."

Haldane showed relatively little interest in heterosis. Haldane (1954) later commented that selection would favor duplicating a heterotic locus, so that every individual, not just the fortunate heterozygotes, could posses both alleles. Perhaps the most familiar example of heterosis is the one in West African blacks for sickle cell anemia (Haldane 1959; Lewontin 1974). Homozygotes for the sickle cell gene die early and hardly ever reproduce, but heterozygotes are more resistant to malaria than are wild-type homozygotes. Another example is the phosphoglucose isomerase locus in sulphur butterflies (Watt 1983): butterflies that are heterozygous at this locus can fly under a wider variety of conditions than homozygotes. Relatively few heterotic loci are known, however (Lewontin 1974), while gene duplication seems to be a fundamental process in evolution (Ohta 1988). Perhaps Haldane was right.

Now let us turn to the balance between selection and recurrent harmful mutation. Consider a locus with two alleles, A and a, in a diploid population where successive generations are distinct. Let the relative fitnesses of AA, Aa, and aa be 1, $1 - hK$, and $1 - K$, respectively. Then, in the absence of mutation,

$$\bar{W}/W_{AA} = q^2 + 2(1 - hK)q(1 - q) + (1 - K)(1 - q)^2, \quad (5.3)$$

$$\partial \ln \bar{W}/\partial q = \frac{2[K(1 - h)(1 - q) + Khq]}{1 - hKq(1 - q) - K(1 - q)^2} = 2s(q), \quad (5.4)$$

$$q(n + 1) - q(n) = q(n)[1 - q(n)]s[q(n)]. \quad (5.5)$$

If K and Kh are far smaller than 1, then

$$q(n + 1) - q(n)$$
$$= q(n)[1 - q(n)]\{K(1 - h)[1 - q(n)] + Khq(n)\}. \quad (5.6)$$

Suppose now that a fraction u of the A genes mutate to a each generation. Then

$$q(n + 1) - q(n)$$
$$= q(n)[1 - q(n)]\{K(1 - h)[1 - q(n)] + Khq(n)\} - uq(n). \quad (5.7)$$

When $q(n + 1) = q(n)$,

$$(1 - q)[K(1 - h)(1 - q) + Khq] = u, \quad (5.8)$$

at which point the number of a genes eliminated by selection balances those appearing by mutation. If most a genes are eliminated as heterozygotes, then Khq greatly exceeds $K(1 - h)(1 - q)$. Moreover, if Kh also greatly exceeds the mutation rate u, then q will be nearly 1, and $u = (1 - q)Kh$. In this case, the number of mutations occurring balances the number eliminated in heterozygotes. Here, a's frequency will be

$$1 - q = u/Kh. \tag{5.9}$$

If $h = 0$, and a is completely recessive, then $K(1 - q)^2 = u$, and a's frequency will be

$$1 - q = \sqrt{u/K}. \tag{5.10}$$

The more nearly recessive a is, that is to say, the lower h is, the higher a's frequency will be. If most a-genes are eliminated as heterozygotes, an A-gene is eliminated with almost every a-gene, and a proportion $2u$ of the population is eliminated each generation as a result of selection against a. If a is entirely recessive, selection against it eliminates a fraction u of the population each generation.

Haldane (1932, p. 109) believed that most genetic variation is maintained by the balance between mutation and selection, and that many, perhaps most, individuals are heterozygous for at least one harmful recessive. This view is quite reasonable (Crow and Kimura 1970, pp. 73–77; Simmons and Crow 1977). Mating with a close relative increases the chances that offspring of such matings will be homozygous for a harmful recessive allele. Inbreeding depression, the decreased fertility of matings between close relatives, is generally ascribed to the abundance of loci with rare, recessive mutants. Indeed, the total damage the recessive alleles in a population would inflict if instantly made homozygous has been estimated from the decreased number of children surviving to adulthood from marriages between relatives of different degree (Morton, Crow, and Muller 1956).

Finally, how much variation can be maintained by neutral mutations? Consider a single locus, with alleles A and a, in a population of sexual haploids, where successive generations are distinct, and each generation contains exactly N mature individuals: call individual x the offspring of individual y if x inherits its allele at this locus from individual y. In the absence of mutation, A's frequency changes from generation to generation: simply by chance, some individuals leave no offspring, while others leave several. Even if A and a are equally fit, the descendants of all but one individual die out, simply as a result of genetic drift. How quickly does this occur? What does this tell us about the amount of genetic variation that can be maintained by neutral mutations (mutations with no effect on fitness)?

To find out, we use a variant of Wright's coefficient of inbreeding. Suppose that, at generation 0, a species of "infinite" extent is divided into populations of size N, and suppose that, forever after, individuals mate at random within their subpopulations. Let $F(n)$ be the probability that, at generation n, two genes sampled at random from the same subpopulation are descended from the same gene at generation 0. If A's frequency in the species as a whole is q, the frequency of Aa heterozygotes in the species at generation n is $2[1 - F(n)]q(1 - q)$: here, $F(n)$ is the coefficient of inbreeding for the species, defined in terms of the probability that genes carried by uniting gametes are "identical by descent." Within each subpopulation, the descendants of a single one of generation 0's genes take over, so F increases toward 1. The speed of F's increase measures the impact of "genetic drift," the changes in A's frequency due to random differences in the reproductive success of different individuals.

Now let us focus on a single subpopulation. Let $F(n)$ be the probability that two individuals sampled at random from generation n have the same allele at the A locus. To calculate $F(n + 1)$ from $F(n)$, suppose that the N mature individuals of generation n make equal (and large) numbers of gametes. Sample N genes at the A locus at random from these gametes to pass on to the next generation's adults. Then the probability that two individuals of generation $n + 1$ acquired their gene at the A locus from the same parent is $1/N$. The probability is $(1 - 1/N)F(n)$ that these two individuals are descended from different carriers of the same A-locus allele. Thus (Crow and Kimura 1970, p. 101),

$$F(n + 1) = 1/N + (1 - 1/N)F(n), \tag{5.11}$$

$$1 - F(n + 1) = (1 - 1/N)[1 - F(n)]. \tag{5.12}$$

If our subpopulation contains N mature *diploids* each generation, which mate at random, and if we now let $F(n)$ be the probability that two genes at the A locus sampled at random from the mature individuals of generation n are the same allele, then

$$1 - F(n + 1) \approx (1 - 1/2N)[1 - F(n)], \tag{5.13}$$

regardless of whether the population is hermaphroditic, or consists of separate sexes in equal numbers (Wright 1931; Crow and Kimura 1970, pp. 102–105).

If there are only two alleles at this locus, and if, at generation n, A's frequency in the subpopulation is $q'(n)$, then on average

$$2q(1-q)[1-F(n)] = 2q'(n)[1 - q'(n)]. \tag{5.14}$$

According to equations (4a.19) and (4b.21), $q(1 - q)$ is very nearly proportional to the subpopulation's genic variance at this locus. If so, then in a diploid population with N reproductive adults each generation, the chances of sampling the $2N$ genes of generation $n + 1$'s adults from the gametes of generation n reduces the genic variance by the factor $1 - 1/2N$ each generation. This result applies generally for populations where successive generations are distinct (Fisher 1930a; Lande 1976), as we shall see later. The total genic variance contributed to this subpopulation by the descendants of a single neutral mutation therefore averages

$$1 + (1 - 1/2N) + (1 - 1/2N)^2 + \cdots = 1/[1 - (1 - 1/2N)]$$
$$= 2N \qquad (5.15)$$

times that from the newly mutant individual itself. Here, we have used the formula

$$(1 + a + a^2 + a^3 + \cdots + a^{n-1})(1 - a) = 1 - a^n \qquad (5.16)$$

to conclude that, if a^2 is less than 1, then

$$1 + a + a^2 + a^3 + \cdots + a^n + a^{n+1} + \cdots = 1/(1 - a). \quad (5.17)$$

For comparison, consider once again a mutant allele a whose heterozygotes have selective advantage Kh. If most mutants are eliminated as heterozygotes, then equation (5.9) implies that a's frequency is u/Kh; that is to say, there are $1/Kh$ a's in the population for each new mutant a. It is as if, in the absence of mutation, selection reduces a's frequency by an average of $1 - Kh$ per generation. If so, the total number of a's descended from a single new mutant would average

$$1 + (1 - Kh) + (1 - Kh)^2 + (1 - Kh)^3 + \ldots = 1/Kh. \quad (5.18)$$

The genic variance maintained by mutation to a is thus about $1/Kh$ times the variance introduced by mutation each generation. Selection is thus $Kh/(1/2N)$, or $2NKh$, times as effective as chance fluctuations in allele frequencies at eliminating genic variance. Put another way, a neutral mutation's descendants contribute an average of $2NKh$ times as much genic variance to future generations as do those of mutations with heterozygous disadvantage Kh. Thus selection is a stronger influence than random changes in allele frequency when $2NKh$ exceeds 1.

In fruit flies, *Drosophila melanogaster*, "quasinormal" mutations have an average selective disadvantage of 0.027 when homozygous (Simmons and Crow 1977, Table 1). New mutations average 40% as harmful when heterozygous (Simmons and Crow 1977, p. 62), so the disadvantage of such heterozygous mutants averages 0.01. The heritable variation in the number of abdominal bristles of *Drosophila* is roughly a thousand times that introduced by one generation of new mutations (Clayton and Robertson 1955), suggesting that mutations affecting bristle number have an average selective disadvantage of 0.001. The heritable variation at isozyme loci is 10,000 times that introduced by one generation's mutations, as if these mutations suffered an average selective disadvantage of 0.0001. Is heterosis more prevalent in isozymes than in other types of genes? Are isozyme variants more nearly neutral than mutants affecting bristle number or viability, as Kimura (1983a) would have us believe? Does currently "neutral" genetic variation play a role in a population's capacity to adapt to environmental change?

There is a whole class of mechanisms for the maintenance of genetic variation which Haldane did not consider—that class of mechanisms whereby a genotype's fitness is "frequency-dependent", declining as the genotype's frequency increases. The classic example of frequency-dependent fitnesses is the "rare male. advantage" in *Drosophila*: if virgin *Drosophila* females are exposed simultaneously to a mixture of *Drosophila* males of different genotype, they mate most frequently with those of the rarest genotype (Ehrman and Probber 1978, Leonard and Ehrman 1976). As each genotype is most fit when rare, natural selection tends to maintain all of them in a population. Considering the almost inhuman rejection by human beings of the "abnormal" in favor of the normal or the "supernormal", whether as mates or in more mundane interactions, the behavior of these *Drosophila* males is puzzling, to say the least.

On the other hand, the diversity of trees in tropical rainforest, where diseases and insect pests abound, is said to be so astoundingly great because the rarer a species of tree, the less it suffers from specialist insect pests (Janzen 1970, Leigh 1990). Hamilton (1982) suggests a similar explanation for genetic diversity within a population. Pathogens differ in their ability to infect hosts of different genotype (Haldane, p. 76). Hamilton concludes that, just as a species of tropical tree suffers less from its insect pests when rarer and harder to find, so a genotype is more fit, less plagued by its specialized pathogens, when rare. Hamilton hangs weighty conclusions on this explanation, for he uses this explanation to seek an understanding of why females prefer the mates they do, and even to explain the prevalence of sexual reproduction. Thus frequency-dependent fitnesses could become as important to evolutionary theorists as frequency-dependent competition is to ecologists.

6 THE EVOLUTION OF DOMINANCE (MUTATION PRESSURE AS A CAUSE OF EVOLUTION)

In section 5 we saw how, in a diploid locus with alleles A and a, mutation that causes a proportion u of each generation's A-genes to be replicated as a, where a has selective disadvantage Kh when heterozygous; this recurrent mutation leads to the selective elimination of a fraction $2u$ of the population each generation.

Now consider a locus with alleles B and b, where B makes A dominant over a without affecting aa individuals. Thus $AaBB$ and $AaBb$ have the same fitness as AA. B thereby spares a fraction $2u$ of its bearers the early death they would otherwise have suffered as a result of selection against Aa. B accordingly derives selective advantage $2u$ from "modifying" Aa so as to make A dominant over a. As mutation rates are very low, usually 10^{-5} or less, the selective differential involved is minute.

Fisher (1930) argued that the dominance of normal or "wild-type" alleles over their mutant alternatives, a circumstance that Haldane recognized as normal, is an evolved phenomenon. Fisher (1935, 1938, 1958a) showed that artificial selection readily modified the dominance of one allele over another. He asserted that the evolution of dominance demonstrated the precision of adaptation and the efficacy of minute but long-lasting selective differentials. Wright (1934) believed, on the other hand, that since genes catalyze chemical reactions, they affect the expression of all other genes. Therefore an allele that modified the degree of dominance at another locus would have many other effects, whose influence on fitness would swamp the minute selective advantage of modifying dominance. Wright's view, that dominance is a physiological phenomenon, which Haldane seems to have shared, became the consensus of evolutionary theory (Kacser and Burns 1981).

If dominance is, at least in part, a consequence of the biochemical organization of living things, then diploidy—doubling the representation of each chromosome—would benefit a population of sexual haploids in the short run (cf. Haldane 1932, p. 109), for then most mutant alleles would be masked by dominant counterparts. After the recessive mutants have spread in response to the reduction of their selective disadvantage, increasing the number of sets of chromosomes yet again (polyploidy) would confer the same benefit. This may explain why polyploidy is so important a mode of speciation in plants. The primary advantage of diploidy, however, is probably the protection it offers against "somatic" mutations, mutations that occur during cell division as an organism grows (Crow and Kimura 1970, p. 316). This may explain why haploids are rarer among larger and more complex organisms.

7 INCOMPATIBLE TRAITS

Sometimes organisms must "choose" between incompatible aptitudes. Fruit flies that reproduce early are shorter-lived than those that do not: fruit flies face a trade-off between early or rapid reproduction and long life (Rose and Charlesworth 1981a,b). Preference for patchy, ephemeral resources, where colonists should multiply rapidly in hopes that at least one descendant will find a new patch when the old is exhausted, is compatible with early, rapid reproduction. Late reproduction and long generation times are. incompatible with dependence on ephemeral resources, but quite suitable for exploiters of a stable food supply. Enzymes may face a trade-off between effective function when cool and preservation of form and function (stability) when warm. The phospho-glucose isomerases that help butterflies mobilize energy for quick take-offs and sustained flights face such trade-offs (Watt 1983, pp. 712–714). Accordingly, an allele that causes its bearers to prefer cool sites is incompatible with an allele that programs enzymes which function best in hot weather, and vice versa: butterflies are more fit when they prefer the sites in which they function best.

To consider how a population confronted with such a trade-off might evolve, Haldane (1932, pp. 107–109) considered the effects of natural selection acting simultaneously at two loci. Let us consider, then, two loci, one with alleles A, a, the other with alleles B, b, in a population of sexual haploids. Being haploid, an individual has only one allele at each locus, which it has equal chance of inheriting from its mother or its father. The probability is $1 - r$ that it inherits its allele at both loci from the same parent. If $r = 1/2$, alleles at the two loci are said to segregate, or assort, independently, as happens when the loci are on different chromosomes. If the two loci occur on the same chromosome, they are said to be "linked." Here, r is usually less than 1/2, so, more often than not, two alleles from the same chromosome tend to be inherited together. Nonparental genotypes are called "recombinants," and the process by which chromosomes exchange homologous parts, creating such genotypes, is called "recombination."

Let the frequencies of AB, Ab, aB, and ab genotypes among the mature individuals of generation n be x_1, x_2, x_3, and x_4, respectively. If these individuals mate at random, then the frequencies x_1', x_2', x_3', and x_4', of these genotypes among the zygotes of generation $n + 1$ are given in table 7.1, which also shows the contribution to these frequencies from each type of mating. For example, $x_1' = x_1 - r(x_1x_4 - x_2x_3)$.

Let $q = x_1 + x_2$ be the frequency of A, and $p = x_1 + x_3$ be the frequency of B. Then $x_1' + x_2' = x_1 + x_2$: recombination does not alter allele frequencies. Since $x_1 + x_2 + x_3 + x_4 = 1$,

$$x_1 - pq = x_1(x_1 + x_2 + x_3 + x_4) - (x_1 + x_2)(x_1 + x_3)$$
$$= x_1x_4 - x_2x_3, \tag{7.1}$$

$$x_1 = pq + D, D = x_1x_4 - x_2x_3. \tag{7.2}$$

Since $x_1 + x_2 = q$, $x_2 = (1 - p)q - D$; since $x_1 + x_3 = p$, $x_3 = p(1 - q) - D$, and $x_4 = (1 - p)(1 - q) + D$. D is called "linkage disequilibrium," and it represents the covariance of A and B among the individuals of this population.

The linkage disequilibrium D' among the zygotes of generation $n + 1$ is

$$D' = x_1' - pq = x_1 - rD - pq$$
$$= pq + D - rD - pq = D(1 - r). \tag{7.3}$$

In the absence of natural selection, linkage disequilibrium diminishes by geometric progression:

$$D(n) = (1 - r)D(n - 1) = (1 - r)^n D(0). \tag{7.4}$$

Now suppose that these genotypes differ in fitness. Suppose, for example, that the allele A renders its bearers distasteful, while b confers bright "warning" colors on its bearers, and B disguises its bearers against their preferred background. Although some palatable animals have mimicked bright, distasteful models, we assume that, under our circumstances, bright palatable animals are less fit than bright, distasteful ones or palatable, disguised ones. Finally, if

TABLE 7.1 *Joint Changes of Allele Frequencies at Two Loci in a Population of Sexual Haploids*

Frequencies of maternal genotypes	AB x_1	Ab x_2	aB x_3	ab x_4
		Frequencies of paternal genotypes		
$AB\ x_1$	$x_1^2 AB$	$\frac{1}{2}x_1x_2 AB$ $\frac{1}{2}x_1x_2 Ab$	$\frac{1}{2}x_1x_3 AB$ $\frac{1}{2}x_1x_3 aB$	$\frac{1}{2}(1-r)x_1x_4 AB$ $\frac{1}{2}rx_1x_4 Ab$ $\frac{1}{2}rx_1x_4 aB$ $\frac{1}{2}(1-r)x_1x_4 ab$
$Ab\ x_2$	$\frac{1}{2}x_2x_1 AB$ $\frac{1}{2}x_2x_1 Ab$	$x_2^2 Ab$	$\frac{1}{2}rx_2x_3 AB$ $\frac{1}{2}(1-r)x_2x_3 Ab$ $\frac{1}{2}(1-r)x_2x_3 aB$ $\frac{1}{2}rx_2x_3 ab$	$\frac{1}{2}x_2x_4 Ab$ $\frac{1}{2}x_2x_4 ab$
$aB\ x_3$	$\frac{1}{2}x_3x_1 AB$ $\frac{1}{2}x_3x_1 aB$	$\frac{1}{2}rx_3x_2 AB$ $\frac{1}{2}(1-r)x_3x_2 Ab$ $\frac{1}{2}(1-r)x_3x_2 aB$ $\frac{1}{2}rx_3x_2 ab$	$x_3^2 aB$	$\frac{1}{2}x_3x_4 aB$ $\frac{1}{2}x_3x_4 ab$
$ab\ x_4$	$\frac{1}{2}(1-r)x_4x_1 AB$ $\frac{1}{2}rx_4x_1 Ab$ $\frac{1}{2}rx_4x_1 aB$ $\frac{1}{2}(1-r)x_4x_1 ab$	$\frac{1}{2}x_4x_2 Ab$ $\frac{1}{2}x_4x_2 ab$	$\frac{1}{2}x_4x_3 aB$ $\frac{1}{2}x_4x_3 ab$	$x_4^2 ab$

NOTES: At the intersection of the row appropriate to the mother's genotype and the column appropriate to the father's genotype are the frequencies of zygotes of different genotypes contributed to the population at large by matings of this type. We thereby find that the zygotic frequencies are:

$$x_1' = x_1x_1 + x_1x_2 + x_1x_3 + x_1x_4(1-r) + rx_2x_3 = x_1 - r(x_1x_4 - x_1x_2)$$
$$x_2' = x_1x_2 + x_2x_2 + x_3x_2(1-r) + rx_4x_1 + x_4x_2 = x_2 + r(x_1x_4 - x_2x_3)$$
$$x_3' = x_3x_1 + x_3x_2(1-r) + rx_1x_4 + x_3x_3 + x_3x_4 = x_3 + r(x_1x_4 - x_2x_3)$$
$$x_4' = x_4x_1(1-r) + rx_2x_3 + x_4x_2 + x_4x_3 + x_4x_4 = x_4 - r(x_1x_4 - x_2x_3).$$

the biochemistry of distastefulness renders distasteful individuals less efficient, and if B's disguise is sufficient protection, then disguised, distasteful animals will be less fit than disguised, palatable ones. Therefore, let us assume that the proportions of AB, Ab, aB, and ab zygotes surviving to maturity are $1 - k$, 1, 1, and $1 - k$, respectively. Then AB's frequency among the adults of generation $n + 1$ is

$$x_1(n + 1) = x_1'(1 - k)/[x_1'(1 - k) + x_2' + x_3' + x_4'(1 - k)]. \quad (7.5)$$

Setting $x_1' + x_2' + x_3' + x_4' = 1$, $x_1' = x_1 - rD$, and so forth, we find that

$$x_1(n + 1) = \frac{(x_1 - rD)(1 - k)}{1 - k(x_1 + x_4 - 2rD)}$$

$$= x_1 - rD + \frac{k(x_1 - rD)[x_1 + x_4 - 2rD - 1]}{1 - k(x_1 + x_4 - 2rD)}, \quad (7.6)$$

$$x_2(n + 1) = \frac{x_2 + rD}{1 - k(x_1 + x_4 - 2rD)}$$

$$= x_2 + rD + \frac{k(x_2 + rD)[x_1 + x_4 - 2rD]}{1 - k(x_1 + x_4 - 2rD)}, \quad (7.7)$$

$$x_3(n + 1) = \frac{x_3 + rD}{1 - k(x_1 + x_4 - 2rD)}$$

$$= x_3 + rD + \frac{k(x_3 + rD)(x_1 + x_4 - 2rD)}{1 - k(x_1 + x_4 - 2rD)}, \quad (7.8)$$

Setting $x_1 + x_2 = q$, $x_1 = pq + D$, $x_2 = p(1 - q) - D$, etc., and simplifying,

$$q(n + 1) = q(n) + \frac{k[q(1 - q)(1 - 2p) + (2q - 1)D(1 - r)]}{1 - k[1 - p - q + 2pq + 2D(1 - r)]}, \quad (7.9)$$

$$p(n + 1) = p(n) + \frac{k[p(1 - p)(1 - 2q) + (2p - 1)D(1 - r)]}{1 - k[1 - p - q + 2pq + 2D(1 - r)]}. \quad (7.10)$$

On page 108, Haldane (1932) implicitly assumes that r greatly exceeds the selective differential k, so that D may be set equal to 0 without seriously altering the problem. Following his lead, we find that

$$\frac{dq}{dn} = kq(1 - q)(1 - 2p), \quad \frac{dp}{dn} = kp(1 - p)(1 - 2q). \quad (7.11)$$

Letting $u = q/(1 - q), q = u/(1 + u); v = p/(1 - p), p = v/(1 + v)$; and noticing that $dq/dn = Kq(1 - q)$ implies $du/dn = Ku$ even when K depends on q, we find

$$\frac{du}{dn} = ku(1 - 2p) = ku\left(\frac{1 - v}{1 + v}\right); \quad \frac{dv}{dn} = kv(1 - 2q) = kv\left(\frac{1 - u}{1 + u}\right).$$

$$(7.12)$$

We may learn what we need from these equations by graphical methods (phase plane analysis). Equations (7.12) tell us that u only increases if v is less than 1, v only increases if u is less than 1. Using this fact, we may separate the plane of possible sets of values of u, v into four quadrants (figure 7.1). If u and v are both less than 1, they both increase; if they both exceed 1, they both decrease; if only u exceeds 1, u increases and v decreases; if only v exceeds 1, v increases and u decreases. Where $v = 1$, $du/dn = 0$, so that the curve $u(n), v(n)$ must cross the line $v = 1$ horizontally, while, where $u = 1$, $dv/dt = 0$, and the solution curve must move vertically. These constraints ensure that either A or B can spread through the population, but not both (figure 7.1).

To establish formally which of the two wins, divide du/dn by dv/dn to obtain

$$\frac{du}{dv} = \frac{u(1 - v)(1 + u)}{v(1 - u)(1 + v)}; \quad \frac{(1 - u)du}{u(1 + u)} = \frac{(1 - v)dv}{v(1 + v)}. \quad (7.13)$$

Since $(1 - u)/u(1 + u) = 1/u - 2/(1 + u)$, $du/u = d \ln u$, and similarly for the term in v, we may rewrite equation (7.13) as

$$d \ln u - 2d \ln (1 + u) = d \ln v - 2d \ln (1 + v) \quad (7.14)$$

$$u(n)[1 + u(0)]^2/u(0)[1 + u(n)]^2$$

$$= v(n)[1 + v(0)]^2/v(0)[1 + v(n)]^2 \quad (7.15)$$

$$\frac{u(n)}{v(n)} = \frac{u(0)}{v(0)} \frac{[1 + v(0)]^2}{[1 + u(0)]^2} \frac{[1 + u(n)]^2}{[1 + v(n)]^2}. \quad (7.16)$$

Suppose that $u(0)$ and $v(0)$ are both less than 1, but that $u(0)$

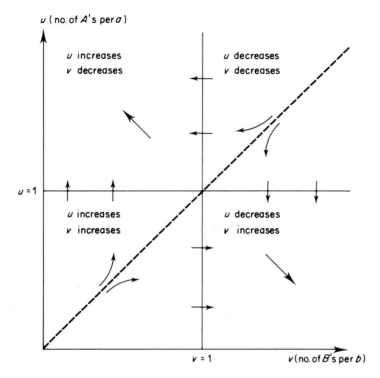

Figure 7.1. "Phase plane" analysis of changes of allele frequencies at two loci. A solution to equations (7.12) for u and v is a "solution curve" $u(n)$, $v(n)$. If this curve enters the lower right-hand quadrant, it can never get out: v must increase indefinitely, and u decline to 0. Similarly, the upper left-hand quadrant is a trap, where u must increase indefinitely and v decline to 0. The diagonal dashed line is the "separatrix." If u exceeds v to begin with, so $u(0)$, $v(0)$ lies above the diagonal, A is the eventual winner and B must disappear, whereas if v exceeds u, it is B which wins.

exceeds $v(0)$. Then, to begin with, they will both increase. However, the ratio u/v also increases, for the factor multiplying $u(0)/v(0)$, which is 1 at generation 0, must progressively increase as u and v grow.

In practice, such incompatible genotypes are suited to different resources or different habitats. In this case, two reproductively isolated populations, each containing one of the two incompatible genotypes, could coexist, for each population's numbers would be limited by a different set of factors.

Given such a potential for speciation, how can one species become two? The first requirement is that one genotype not swamp the other before speciation can begin. This condition is met if the numbers of each genotype are limited by different resources, so that the population will continue to maintain both genotypes. What happens then?

Felsenstein (1981) considers a population divided equally between two environments; environment I which favors A over a, and environment II which favors B over b. Here, numbers of individuals maturing in each environment are limited separately: the same number per generation mature in each. Following Felsenstein, let us first assume that half the offspring of the individuals in each environment migrate to the other shortly after birth. Let the relative fitnesses of the various genotypes in each environment, and their average fitnesses for the population as a whole, be as in table 7.2.

Now consider a third locus, with alleles C and c, where a proportion d of the C-bearers mate only with other C's, and a proportion d of the c-bearers only with other c's, while the remainder mate at random. If one of the mating-preference alleles predominates among the Ab's, and the other among the aB's, there will be partial reproductive isolation between Ab's and aB's, which Felsenstein interprets as progress toward speciation. Felsenstein asked what circumstances would allow such reproductive isolation to evolve. If the proportion of offspring exchanged between each environment is $1/2$, then, if $s = 1$, d must be 0.67 if the mating preference alleles are to become associated with the incompatible genotypes Ab and aB. If the proportion of offspring exchanged between each environment is $1/10$, and $s = 1$, a value d of 0.47 would allow mating preferences to align with these incompatible genotypes. Lowering the exchange of migrants between the two populations facilitates speciation. Nevertheless, Felsenstein's

TABLE 7.2

	Environment I	Environment II	Average
Ab	$(1 + s)^2$	1	$1 + s + \frac{1}{2}s^2$
AB	$1 + s$	$1 + s$	$1 + s$
ab	$1 + s$	$1 + s$	$1 + s$
aB	1	$(1 + s)^2$	$1 + s + \frac{1}{2}s^2$

model suggests that genotypes must indeed be very incompatible, and that these genotypes must be associated with alleles governing very discriminating mating preferences, as if driven by a very strong sexual selection, if the one species is to become two.

Wilson and Turelli (1986) imagined a locus with two alleles, say A and a, in a population of diploid predators where AA exploit prey species 1 more effectively than prey species 2, while the reverse is true for aa. Aa are phenotypically of intermediate dominance: we may assume they feed equally well on both prey. If, as a result, the heterozygotes feed relatively inefficiently on both kinds of prey, then Aa are less fit than AA or aa, because the abilities to exploit these two kinds of prey are incompatible. Even so, A and a will persist in the population, because the numbers of AA and aa are limited by different factors. Such a single-locus model with two incompatible alleles would provide a basis for a study of speciation analogous to Felsenstein's.

Speciation is one of the mysteries of evolutionary biology. Biologists are beginning to realize how little they understand about the process. Some disagreements about speciation are as old as evolutionary theory itself. Darwin (1859) thought that reproductive isolation evolved as an automatic byproduct of the evolutionary divergence of populations in different environments. Wallace, on the other hand, thought that natural selection would favor reproductive isolation if hybrid matings between members of adjacent populations undergoing evolutionary divergence yielded fewer, or inferior, offspring (Mayr 1963, p. 548). Felsenstein's model suggests that selection for reproductive isolation is effective only under extreme circumstances.Nevertheless, Coyne and Orr (1989) present convincing evidence that selection for discriminatory mating preferences has played a distinct role in the evolution of reproductive isolation between overlapping populations of *Drosophila*.

8 SELECTION ON A QUANTITATIVE (METRICAL) TRAIT DETERMINED BY MANY GENES

Natural selection is usually discussed as an agent of evolutionary change. In general, however, most of the characteristics of a species' individuals are well suited to their way of life. For these characteristics, natural selection is a stabilizing influence that eliminates deviant genotypes (Haldane 1959, pp. 120ff). As an example, consider a quantitative characteristic (Haldane and Fisher called them "metrical characters") determined by many genes. Suppose that, like weight or length, this characteristic can be specified by a single number M. Let the optimum value (optimum measure) for this characteristic be M^*, and let the relative fitness of an individual with value M be $\exp - s(M - M^*)^2$. How does selection affect the frequencies of alleles influencing this characteristic? How much genetic variation in this characteristic will selection allow?

These two questions together constitute one of the "classic problems" of population genetics. A surprising number of distinguished theorists have tackled it (Fisher 1930a; Haldane 1932; Wright 1935; Kimura 1965; Lande 1976b; Felsenstein 1977; Bulmer 1980; Turelli 1984; Barton 1986a, 1989; Slatkin 1987; Keightley and Hill 1988; Bürger 1988; Nagylaki 1989a). This list is by no means exhaustive. Many of the cognoscenti would be justly shocked by the papers I omitted from this hasty list. This problem has left more than one theorist in the lurch, dependent for rescue on Lerch's zeta (Turelli 1984).

Before launching into the theory, let us consider some examples of selection on quantitative characteristics. First, how strong is the selection likely to be? Karn and Penrose (1951, cited in Haldane 1959) recorded the birth weights of 13,700 babies born in London between 1935 and 1946. Only 1.8% of the babies with birth weight between 7.5 and 8 pounds died within 28 days of birth, while 4.5% of all the babies did so. Stabilizing selection for birth weight (or its correlates) thus imposed a mortality of $4.5 - 1.8$, or 2.7% on this population. Shami and Tahir (1979) found that, in a town near Lahore in Pakistan, completed families where the average of mother's and father's heights was between 158 and 159 cm averaged 3.58 living children apiece, while completed families averaged 3.41 living children apiece in the town at large, 0.17 child, or 4.7%, less. Thus stabilizing selection on human height (or its correlates) imposed a mortality of 4.7% on this population. Turelli (1984, p. 185) summarizes information suggesting that a typical stabilizing selection imposes a mortality of about 5%. Anywhere between 10 and 1,000 loci may affect a given quantitative characteristic (Turelli 1984, p. 175).

Now to the theory. Let us assume, to be specific, that our characteristic is governed by m loci, each with two alleles apiece, in a population of randomly mating diploids. Suppose that the optimum value for our characteristic is 0 (subtract M^* from all measurements, if necessary, to make this true). Let each locus i have two alleles, A_i and a_i; let each of an individual's A_i genes add an amount b to the value of the characteristic, while each a_i adds $-b$; and let A_i's frequency be q_i. Since an individual carries two genes at each locus, the average value of M for the population as a whole will be

$$\bar{M} = 2 \sum_{i=1}^{m} [q_i b - (1 - q_i)b] = 2b \sum_{i=1}^{m} (2q_i - 1). \quad (8.1)$$

Assume, moreover, that alleles at different loci are distributed independently: a dangerous assumption, but not misleading under these circumstances (Turelli and Barton 1990). Then the genic variation in the population as a whole is the sum of the genic variances from each locus. Since an allele is distributed independently among maternal and paternal chromosomes, the genic variance V_i from locus i is twice the variance from its maternal chromosomes. The genic variance $\frac{1}{2}V_i$ from the maternal chromosome of locus i is

$$q_i[b - b(2q_i - 1)]^2 + (1 - q_i)[-b - b(2q_i - 1)]^2$$
$$= 4b^2[q_i(1 - q_i)^2 + (1 - q_i)q_i^2] = 4b^2 q_i(1 - q_i). \quad (8.2)$$

Then the genic variance for the population as a whole is

$$V_g = \sum_{i=1}^{m} V_i = 8b^2 \sum_{i=1}^{m} q_i(1 - q_i). \quad (8.3)$$

Now let us express the mean value $\bar{M}$ of this characteristic as the contribution $2b(2q_1 - 1)$ from locus 1 plus the contribution M' from the remaining loci. Similarly, the genic variance V_g is the contribution $8b^2 q_1(1 - q_1)$ from locus 1 plus the contribution V_g' from the remaining loci. Thus

$$\bar{M} = 2b(2q_1 - 1) + M', \quad (8.4)$$

$$V_g = 8b^2 q_1(1 - q_1) + V_g'. \quad (8.5)$$

Finally, let the total phenotypic variance in this characteristic be $V = V_g + V_e$, where V_e represents the contribution of random environmental effects—whose magnitude is assumed independent of genotype—to this variance.

Let the proportion of individuals with trait values between M and $M + dM$ be

$$f(M)dM = (dM/\sqrt{2\pi V}) \exp - (M - \bar{M})^2/2V. \qquad (8.6)$$

Then, since the fitness of an individual whose trait measures M is

$$W(M) = \exp - sM^2, \qquad (8.7)$$

the mean fitness for the population as a whole is

$$\bar{W} = \int_{-\infty}^{\infty} dMf(M)W(M), \qquad (8.8)$$

$$\bar{W}\sqrt{2\pi V} = \int_{-\infty}^{\infty} dM \exp - \left[sM^2 + \frac{(M - \bar{M})^2}{2V} \right]$$

$$= \int_{-\infty}^{\infty} dM \exp - \frac{2VsM^2 + (M - \bar{M})^2}{2V}$$

$$= \int_{-\infty}^{\infty} dM \exp - \frac{M^2 - 2M\bar{M}/(1 + 2Vs) + \bar{M}^2/(1 + 2Vs)}{2V/(1 + 2Vs)}$$

$$= \exp \frac{\bar{M}^2/(1 + 2Vs)^2 - \bar{M}^2/(1 + 2Vs)}{2V/(1 + 2Vs)}$$

$$\times \int_{-\infty}^{\infty} dM \exp - \frac{[M - \bar{M}/(1 + 2Vs)]^2}{2V/(1 + 2Vs)}. \qquad (8.9)$$

Multiplying numerator and denominator of the first exponential by $(1 + 2Vs)^2$, we may express it as

$$\exp \frac{\bar{M}^2 - \bar{M}^2(1 + 2Vs)}{2V(1 + 2Vs)} = \exp - \frac{s\bar{M}^2}{1 + 2Vs}. \qquad (8.10)$$

The integral yields $\sqrt{2\pi V/(1 + 2Vs)}$. Therefore

$$\bar{W} = (1 + 2Vs)^{-\frac{1}{2}} \exp - \frac{s\bar{M}^2}{1 + 2Vs}, \qquad (8.11)$$

$$\ln \bar{W} = -\tfrac{1}{2} \ln (1 + 2Vs) - s\bar{M}^2/(1 + 2Vs). \qquad (8.12)$$

Equation (8.11) implies that when $\bar{M}$ has attained its optimum value 0, then $\bar{W} = (1 + 2Vs)^{-\frac{1}{2}}$. If $2Vs$ is much smaller than 1, then $\bar{W}$ is roughly $1 - Vs$; that is, the mortality imposed by this stabilizing selection is roughly Vs.

Now let us trot out equation (4b.17) to learn how selection affects the frequency of A_1. Since

$$dq_1/dn = \tfrac{1}{2}q_1(1 - q_1) \, \partial \ln \bar{W}/\partial q_1, \tag{8.13}$$

the selective advantage of A_1 over a_1 is $\tfrac{1}{2}\partial \ln \bar{W}/\partial q_1$. A_1 is only favored over a_1 if increasing A_1's frequency increases population fitness, that is, if $\partial \ln \bar{W}/\partial q_1$ is positive. Selection according to this model on a quantitative characteristic cannot explain nonadaptive "orthogenesis."

The selective advantage $\partial \ln \bar{W}/\partial q_1$ of A_1 over a_1 is

$$-[s \, \partial V/\partial q_1 - 2s\bar{M} \, \partial \bar{M}/\partial q_1]/(1 + 2Vs)$$
$$+ 2s^2\bar{M}^2 \, \partial V/\partial q_1/(1 + 2Vs)^2. \tag{8.14}$$

Using equations (8.4) and (8.5), we find that $\partial V/\partial q_1$ is $8b^2(1 - 2q_1)$, $\partial \bar{M}/\partial q_1 = 4b$. Therefore

$$\tfrac{1}{2}\partial \ln \bar{W}/\partial q_1 = [4b^2s(2q_1 - 1) - 4bs\bar{M}]/(1 + 2Vs)$$
$$+ 8s^2\bar{M}^2b^2(1 - 2q_1)/(1 + 2Vs)^2$$
$$= 4b^2s[(2q_1 - 1 - \bar{M}/b)(1 + 2Vs)$$
$$+ s\bar{M}^2(1 - 2q_1)]/(1 + 2Vs)^2. \tag{8.15}$$

In his attempt to explain nonadaptive "orthogenesis," Haldane (1932, pp. 110–112) assumed that $\bar{M}$ was continually equal to 0. Were this true, A_1's selective advantage would be proportional to $2q_1 - 1$. If its frequency exceeded $\tfrac{1}{2}$, its frequency would increase further; were its frequency less than $\tfrac{1}{2}$, it would decrease further. One might conclude that, if selection suddenly stopped just when the frequencies of most such alleles had surpassed $\tfrac{1}{2}$, these alleles would spread to fixation, conferring a nonadaptive momentum on this evolutionary change. $\bar{M}$, however, consists of a series of terms of the form $2b(2q_i - 1)$. Equation (8.15) tells us that it would take no more than one locus fixed the "wrong" way to change $\bar{M}$ from $2b$ to $-2b$, which would be enough to reverse the direction of selection on A_1. Therefore, Haldane's (p. 112) explanation of orthogenesis is erroneous.

When $\bar{M}$ is close to the optimum, V should be far larger than $\bar{M}^2$. Thus $1 + 2Vs$ should be far larger than $s\bar{M}^2$. Moreover, as we have seen, sV is the selective disadvantage of the "average" individual relative to one for which $M = 0$. As alleles at different loci are distributed more or less independently only if sV is weak, we will assume that $1 + 2Vs = 1$. Then equation (8.15) for the selective advantage of A_1 over a_1 can be simplified to read

$$4bs[2q_1 - 1)b - \bar{M}] = 4bs[(1 - 2q_1)b - M'], \qquad (8.16)$$

where, as in equation (8.4), $M' = \bar{M} - 2b(2q_1 - 1)$. This equation implies that half of the m loci influencing the value M of our trait should carry only A_i, and half only a_i. If there is one locus left over (that is, if M' is exactly 0), the remaining locus should be in stable polymorphism (Barton 1986a). Without mutation, selection according to this model maintains almost no genetic variation.

Now suppose that each A_1 mutates to a_i, and vice versa, at the rate u. If u is far smaller than $4b^2s$—which might be true—then, for half the loci, we might expect that $q_i = u/4b^2s$, while, for the other half, $1 - q_i$ has this value. Equation (8.2) implies that if A_i is rare, its contribution to the genic variance in M is roughly $8b^2q_i$. Setting $q_i = u/4b^2s$, we find that

$$V_i = 2u/s. \qquad (8.17)$$

Thus the total genic variance maintained in M by mutation is

$$V_g = 2mu/s. \qquad (8.18)$$

This stabilizing selection eliminates a proportion $2[4b^2s(u/4b^2s)] = 2u$ of the genes at each of our m loci. The proportion of genes eliminated by this stabilizing selection from the population as a whole is thus

$$2mu = sV_g. \qquad (8.19)$$

The coefficient s is often expressed as $1/2V_s$ (not $\frac{1}{2}Vs$!). Then the fitness of an individual with phenotype M is $\exp -M^2/2V_s$, and $V_g = 4muV_s$.

Equation (8.18) is valid for a wide range of population sizes, so long as the mutation rate u is less than $\frac{1}{2}sb^2$ (Barton 1989), that is, so long as $u/4b^2s$ is less than $1/8$. In other words, equation (8.18) is valid if selection is strong enough to keep all but one allele rare at each locus. Is this reasonable assumption? Barton (1986a, p. 215) argues plausibly that if all loci have equal effect, then at each locus the mutation rate u is much smaller than the selective disadvantage $4b^2s$ of the rarer allele. The probability mu that a gamete contains a new mutation affecting our quantitative characteristic is roughly 10^{-2} (Turelli 1984, p. 175). Each generation, mutation increases the phenotypic variance of our quantitative characteristic by an amount $8mub^2$. For those characteristics that have been studied, the variance supplied by one generation of new mutations is 0.1% of the environmental component V_e of the phenotypic variance (Lande 1976b). Finally, for many such characteristics, sV_e is roughly 0.03 (Turelli 1984, pp. 184–185). Thus, if we set $4mub^2s = (4mub^2/V_e)(sV_e)$, then

$$u/4b^2s = (m^2u^2)/m(4mub^2s) = 10^{-4}/m(10^{-3})(0.03) = 3/m, \quad (8.20)$$

where m is the number of loci affecting the characteristic. To bring the mutation rate per locus as low as 10^{-4} (the usual figure for mutation rate per locus is 10^{-6}), m must be 100, in which case the average selective disadvantage of a new mutation affecting a quantitative characteristic is 0.003. Under these circumstances, selection is strong enough to override both mutation and genetic drift.

The argument given above for equation (8.18), however, is only valid if $u(m + 1)^2/4b^2s$ is less than 1. If $u/4b^2s = 3/m$, as suggested in equation (8.20), then $u(m + 1)^2$ is far larger than $4b^2s$. Under these circumstances, each locus has one common and one rare allele (even if m is odd), but, even for even m, there are stable mutation-selection equilibria where the number of loci with rare A-allelesis as high as $\frac{1}{2}[m + (m - 2)\sqrt{u/4b^2s}]$, or as low as $\frac{1}{2}[m - (m - 2)\sqrt{u/4b^2s}]$, as well as for each integer in between. For each of these equilibria, $\bar{M}$ is of absolute value less than b, but the genic variance V_g ranges from $2mu/s$, its value when A is rare at exactly half the loci, to a maximum of $2bm\sqrt{u/s}$, for equilibria where the partitioning between the number of loci with A rare and the number of loci with a rare is as unequal as possible consistent with stability. If $4Nb^2s$ exceeds 2, where N is "effective population size" (roughly the number of reproductive adults per generation: see section 10), genetic drift—chance variations in allele frequency—tends to shuffle the population toward more adaptive equilibria (equilibria with lower genic variance), so that V_g should usually be close to $2mu/s$ (Barton 1989).

Barton's model, however, is dangerously schematic. If alleles at different loci have very different effects on the characteristic, the story becomes much more complicated (Nagylaki 1989). There is every reason to believe, moreover, that the alleles affecting the value of M also affect other characteristics (Turelli 1985). As Turelli (1985) and Turelli and Barton (1990) assure us, the balance between mutation and stabilizing selection on a quantitative trait is a topic that is still replete with mysteries and surprises.

9 EFFECTS OF CHANCE ON ALLELE FREQUENCIES, I. GENERATING FUNCTIONS AND BRANCHING PROCESSES

In section 5 we observed that even when the members of a population produce the same number of equally fit gametes, only a finite sample of these gametes contributes its genes to the adults of the next generation. Simply by chance, therefore, some adults have no offspring and some have several. Thanks to this "sampling error," allele frequencies change somewhat from generation to generation. This process is often called "(random) genetic drift."

Genetic drift is most important for rare alleles. Genetic drift ensures that there is only a small probability that an allele currently represented by a single new mutation will eventually attain the frequency 1, even if that allele has a substantial selective advantage. Here, we use "branching processes" to calculate the probability that an allele currently represented by a single new mutation will eventually spread to fixation. We will also calculate the prospective contribution of such a mutation to genic variance, and its speed of spread or disappearance.

Taking genetic drift into account is a complicated business. Instead of keeping track, say, of the number of A genes per a gene at generation n, we must, if we are concerned with a population of N haploids, keep track, for each generation n, of $N + 1$ probabilities: the probabilities that there are 0 A-genes, 1 A-gene, 2 A-genes, . . . N A-genes in the population at that time. Apparently, taking the effects of chance into account for this population makes our problem $N + 1$ times more difficult. We will need special methods, and special constructs, to attack even the simplest of such problems. The first of our constructs is a "generating function," a device for storing and recovering information about the probabilities of different numbers of descendants (or of different numbers of heads in a series of coin tosses, or whatever).

Consider a locus in a large population of sexual haploids, and call one individual the offspring of another if the one inherits its allele at this locus from the other. Consider an individual bearing the newly mutant allele A at this locus, and suppose she is the only A-bearer in the whole population. How many descendants will she have n generations later? To answer, we first construct a "generating function." If the probability that an A-bearer has r offspring is p_r, then the generating function for the number of her offspring is defined to be

$$f(s) = p_0 + p_1s + p_2s^2 + p_3s^3 + \cdots + p_rs^r \cdots + \qquad (9.1)$$

where s is a "book-keeping variable" with no biological significance. If two A-bearers reproduce independently, the probability that they jointly have r offspring is the probability that the first has none and the second r, plus the probability that the first has 1 and the second $r - 1$, and so forth, or $p_0p_r + p_1p_{r-1} + \cdots + p_rp_0$. This sum, however, is simply the coefficient of s_r in $[f(s)]^2$. The generating function for the number of an A-bearer's grandchildren is therefore

$$f_2(s) = p_0 + p_1 f(s) + p_2[f(s)]^2 + \cdots = f[f(s)]. \qquad (9.2)$$

The generating function $f_n(s)$ for the number of descendants in generation n is defined iteratively:

$$f_1(s) = f(s); f_{n+1}(s) = f_n[f(s)] = f[f_n(s)]. \qquad (9.3)$$

Moreover,

$$\frac{d}{ds} f(s) = f'(s) = p_1 + 2p_2s + 3p_3s^2 + \cdots \qquad (9.4)$$

Therefore $f'(1) = \bar{W}$, the average number of (mature) offspring per parent.

$$\frac{d}{ds} f_n(s) = \frac{d}{ds} f_{n-1}[f(s)] = f'(s)[f'_{n-1}(f)]. \qquad (9.5)$$

If we set $s = 1$, $f(s) = 1 (p_0 = p_1 + \cdots = 1)$, and we find that

$$f'_n(1) = f'_{n-1}(1)f'(1) = \bar{W}f'_{n-1} = \bar{W}^n, \qquad (9.6)$$

which is the expected number of descendants in generation n of our A-bearer. To find the variance in the number of offspring per parent, notice first that the mean square number of offsping per parent is $f''(1) + f'(1)$. Thus the variance v in the number of offspring per parent is

$$v = f''(1) + f'(1) - [f'(1)]^2. \qquad (9.7)$$

A lineage where the ancestor has a random number of offspring, each of which reproduces independently in the same random way, and so forth, is a "branching process." Feller (1968) discusses branching processes in what is the clearest, most thoughtful introduction to probability theory yet written. Harris (1963) treats the topics presented here in far greater generality.

Let us now consider two examples. Suppose first that successive generations are distinct, and that N individuals are sampled at random from the young of generation n to be the adults of the next generation. Consider the bearer of a new mutant A, which has no effect on fitness. What is the generating function $f(s)$ for the number of her offspring? Each sampled young has probability $1/N$ of being hers. As N young are sampled independently,

$$f(s) = [1 - 1/N + s/N]^N. \tag{9.8}$$

The probability p_0 that she has no offspring is $(1 - 1/N)^N$; $p_1 = N(1/N)(1 - 1/N)^{N-1}$. When N is large,

$$f(s) \approx \exp(s - 1)$$
$$= e^{-1}(1 + s + s^2/2 + s^3/6 + \cdots). \tag{9.9}$$

If r is far smaller than N, the probability that she has r offspring is roughly $x^r e^{-1}/r$. Using equations (8.4) and (8.7), we find that the mean and variance in the number of her offspring are

$$\bar{W} = f'(1) = 1;$$
$$v = f''(1) + f'(1) - [f'(1)]^2 = 1. \tag{9.10}$$

Equation (9.8) is the generating function for the number of offspring per parent in the random models of Fisher (1930a), Wright (1931), and Kimura (1964).

Now let generations overlap, and suppose that a mutant A-bearer alive at time t has probability $1/N$ of bearing a young by time $t + 1/N$. Also, let it have probability $1/N$ of dying by then, regardless of whether it has just reproduced, and probability $1 - 1/N$ of living to time $t + 1/N$, in which case its prospects at time $t + 1/N$ are just what they were at time t. Here,

$$f(s) = [1 - 1/N + s/N][1/N + (1 - 1/N)f(s)], \tag{9.11}$$

where the first factor represents her chance of reproducing between times t and $t + 1/N$, and the second, the alternative between dying at time $t + 1/N$, with probability $1/N$, or living to repeat the process, with generating function $(1 - 1/N)f(s)$. Solving for $f(s)$, we find

$$f(s) - \left(1 - \frac{1}{N} + \frac{s}{N}\right)\left(1 - \frac{1}{N}\right)f(s) = \left(1 - \frac{1}{N} + \frac{s}{N}\right)\frac{1}{N}, \quad (9.12)$$

$$f(s) = \frac{N - 1 + s}{2N - 1 - (N - 1)s}. \quad (9.13)$$

When N is large, this generating function is very nearly

$$f(s) = \frac{1}{2 - s} = 1/2 + s/4 + s^2/8 + s^3/16 + \cdots \quad (9.14)$$

In this model, the probability that our A-bearer has r offspring is $p_r = 1/2^{r+1}$. Here,

$$\bar{W} = f'(1) = 1;$$
$$v = f''(1) + f'(1) - [f'(1)]^2 = 2. \quad (9.15)$$

As befits a neutral mutation, $\bar{W} = 1$, but the variance in the number of offspring per parent is twice its value in the Fisher-Wright model where successive generations are distinct. Equation (9.13) is the generating function for the number of offspring per parent in the random models of Moran (Moran 1962; Karlin and McGregor 1962). Moran's model is vastly easier to work with than that of Fisher and Wright. Moreover, with one simple correction, it gives very nearly the same answer.

What is the probability that our newly mutant A-bearer still has A-bearing descendants n generations later? To find out, set $f(s) = 1/(2 - s)$ and calculate $f_n(s)$. Notice that, if

$$f_{n-1}(s) = \frac{n - 1 - (n - 2)s}{n - (n - 1)s}, \quad (9.16)$$

then

$$f_n(s) = f_{n-1}[f(s)]$$

$$= \frac{n - 1 - \dfrac{n - 2}{2 - s}}{n - \dfrac{n - 1}{2 - s}} = \frac{n - (n - 1)s}{n + 1 - ns}. \quad (9.17)$$

Since equation (9.16) applies for $n - 1 = 1$, and since its truth for $n - 1$ implies, by equation (9.17), its truth for n, it follows by induction that

$$f_{n-1}(s) = \frac{n - 1 - (n - 2)s}{n - (n - 1)s}$$

$$= \frac{n - 1}{n} + \frac{1}{n}\left[\frac{s/n}{1 - (n - 1)s/n}\right]. \tag{9.18}$$

The probability that our mutant still survives in generation $n - 1$ is only $1/n$; but, if it does survive, the probability that it has r representatives in generation $n - 1$ is $(n - 1)^{r-1}/n^r$, which is nearly $(1/n)\exp - r/n$; and the probability that it has at least r representatives is roughly

$$\int_r^\infty (1/n) \, dx \exp(-x/n) = \exp -r/n. \tag{9.19}$$

More generally, the probability that a neutral mutation represented by a single individual in generation 0 still survives at generation $n - 1$ is $2/vn$, and if it still survives, the probability that it is represented by at least r individuals is $\exp - 2r/vn$ (Fisher 1930a; Harris 1963). Once again we see how, by chance, many lineages die out while a few spread. Nevertheless, the number of descendants of a single neutral mutation is unlikely to be more than a small multiple of the number of generations since its appearance.

What difference does it make if the fitness of our new mutant differs from 1? To find out, set

$$f(s) = \frac{1}{m + 1 - ms}. \tag{9.20}$$

Then $\bar{W} = f'(1) = m$ and $v = m(1 + m)$. As with equations (9.16) and (9.17), one can check that if

$$f_n(s) = \frac{m^n - 1 - (m^n - m)s}{m^{n+1} - 1 - (m^{n+1} - m)s}, \tag{9.21}$$

then $f_{n+1}(s) = f_n[f(s)]$ and $f_1(s) = f(s)$. Equation (9.21) thus follows by induction. We may express $f_n(s)$ as

$$1 - \frac{m^n(m - 1)}{m^{n+1} - 1} + \frac{\dfrac{m^n(m - 1)}{m^{n+1} - 1}\left(\dfrac{m - 1}{m^{n+1} - 1}\right)s}{\left[1 - \dfrac{(m^{n+1} - m)s}{m^{n+1} - 1}\right]}. \tag{9.22}$$

If m slightly exceeds 1, then the probability P that the mutant survives indefinitely is

$$P = \lim_{n \to \infty} \frac{m^n(m-1)}{m^{n+1}-1} = 1 - 1/m. \tag{9.23}$$

When m is nearly 1, P is nearly $k = m - 1$, the selective advantage of the mutant allele A over its prevalent alternative a. If the mutant A is still represented at generation n, then the probability p_{r+1} that it has $r + 1$ representatives then is

$$\frac{p_{r+1}}{1-p_0} = \frac{(m-1)(m^{n+1}-m)^r}{(m^{n+1}-1)^{r+1}}. \tag{9.24}$$

When m^n is far larger than 1 but far smaller than N (so that the mutant is still rare enough for its numbers to multiply by geometric progression, even though in a population of constant size), then

$$\frac{p^{r+1}}{1-p_0} = \frac{m-1}{m^{n+1}}\left[\frac{1-1/m^n}{1-1/m^{n+1}}\right]^r \approx (k/m^n) \exp{-rk/m^n}. \tag{9.25}$$

The probability that, if the allele survives that long, it is represented by at least r bearers in generation n is $\exp{-rk/m^n}$. Its expected number of representatives at generation n, given that it survives so long, is m^n/k: it is as if the mutant started with $1/k$ bearers at generation 0. Indeed, if one were to calculate the time required to replace a by A, given a newly mutant A whose descendants are destined to spread, one does not go far wrong by using equation (4a.9) to calculate the time required for the gene ratio to shift from $1/kN$ to kN, as if the changes from $1/N$ to $1/kN$ and from kN to N were instantaneous.

If m is slightly less than 1, so that $m = 1 - k$, then

$$f_n(0) = (1 - m^n)/(1 - m^{n+1})$$
$$= (1 - e^{-kn})/[1 - (1 - k)e^{-kn}]. \tag{9.26}$$

If e^{-kn} is far smaller than 1, then

$$f_n(0) \approx 1 - ke^{-kn}, \tag{9.27}$$

suggesting that, for large n, A is far less likely to be represented so long after its occurrence than it would be were it neutral.

How do our results depend on the generating function we chose for the number of offspring per parent? What if we only know $f(1)$,

$f'(1)$, and $f''(1)$? Suppose now that $f'(1) = 1 - k$, where k is far smaller than 1, and that $v = f''(1) - k(1 - k) \approx f''(1)$. Let P be the probability that the mutant persists indefinitely. When the probability $f_n(0)$ of A's extinction reaches equilibrium,

$$1 - P = f_n(0) = f[f_n(0)] = f(1 - P). \tag{9.28}$$

Since $f_n(0)$ increases ever more slowly with n, the smallest positive value of $1 - P$ satisfying this equation is the appropriate solution. Taking the "Taylor expansion" of $f(1 - P)$, we find that

$$1 - P = f(1 - P) \approx f(1) - Pf'(1) + \tfrac{1}{2}P^2 f''(1)$$
$$= 1 - P(1 + k) + \tfrac{1}{2}P^2 v, \tag{9.29}$$

$$P \approx 2k/v. \tag{9.30}$$

The decisive ratio is that of A's selective advantage to the variance in its numbers of offspring per parent. This result agrees with our finding that when $f(s) = 1/(m + 1 - ms)$, P is approximately $k = m - 1$ when m only slightly exceeds 1, for in that case, v is very nearly 2.

Next, we ask how rapidly genic variance decays under the influence of genetic drift. Put another way, suppose that, at a large number of loci, each with alleles A_i and a_i, allele frequencies are changing randomly. If, each generation, one locus loses its last A-allele, and one its last a-allele, how many loci still have both A and a?

To be specific, let b_1 loci have one A apiece, b_2 have 2 A's apiece, and so forth. Let

$$F(s) = b_0 + b_1 s + b_2 s^2 + b_3 s^3 + \cdots \tag{9.31}$$

F is a generating function, but it is not a probability-generating function, because the sum of the b_i's does not equal 1. Assume that successive generations are distinct and that, one generation later, the b_i's are all the same, except for b_0, which is increased by 1. If $f(s)$ is the generating function for the numbers of offspring per parent, then

$$F[f(s)] - F(s) = 1. \tag{9.32}$$

Set $f(s) = 1/(2 - s)$, $s(n + 1) = f_{n+1}(0) = 1/[2 - s(n)]$, and $F(0) = s(0) = 0$. Then, by equation (9.18), $s(n + 1) = f_{n+1}(0) = n/(n + 1)$. We may accordingly write

$$F[s(n)] - F[s(n-1)] = 1,$$

$$F[s(n)] - F[s(0)] = F[s(n)] = n. \tag{9.33}$$

Thus F is the inverse of s: F is the same function of s that n is. If $s = n/(n+1)$, $n = s/(1-s)$, and

$$F(s) = s/(1-s) = s + s^2 + s^3 + s^4 + \cdots \tag{9.34}$$

Let there be N individuals in the population, where N is a very large number. Then, when one locus is losing its last A, and another its last a, each generation, there is, on the average, $b_1 = 1$ segregating locus with n A-alleles for every n between 1 and $N-1$, $N-1$ segregating loci in all. Thus genetic variation will decay, simply by chance, at a steady rate of $1 - 2/N$ per generation. This is twice the rate of decay we found in section 5, but in that model the variance in number of offspring per parent was half as great as in this one.

Now let one new mutant, with no effect on fitness, be introduced each generation. Then the total number of loci with no A's increases by 1, and the number with 1 A decreases by 1, each generation, when genetic drift just balances these mutations. Here,

$$F[f(s)] - F(s) = 1 - s. \tag{9.35}$$

Once again, set $s(0) = 0$, $s(n+1) = 1/[2 - s(n)]$, and $F(0) = 0$. We find that

$$F[s(1)] - F[s(0)] = 1 - s(0); \quad F[s(1)] = F(\tfrac{1}{2}) = 1, \tag{9.36}$$

$$F[s(2)] - F[s(1)] = 1 - s(1) = \tfrac{1}{2};$$
$$F[s(2)] = F(2/3) = 1 + \tfrac{1}{2}, \tag{9.37}$$

$$F[s(n)] = 1 + 1/2 + 1/3 + \cdots$$
$$+ 1/n \approx \ln n \approx -\ln(1/n). \tag{9.38}$$

Since $s(n) = n/(n+1)$, $1/n = 1 - s(n-1)$. Therefore

$$F[s(n)] \approx -\ln(1/n) = -\ln[1 - s(n-1)]. \tag{9.39}$$

We are concerned with $F(s)$ as it approaches steady state; that is, as $s(n)$ approaches 1 and $s(n) - s(n-1)$ becomes small compared with $1 - s(n)$. Then

$$F(s) = -\ln(1-s). \tag{9.40}$$

Since

$$1/(1 - x) = 1 + x + x^2 + \cdots, \int \frac{dx}{1 - x}$$

$$= -\int d \ln (1 - x) = -\ln (1 - x), \qquad (9.41)$$

$$\int \frac{dx}{1 - x} = \int dx[1 + x + x^2 + \cdots]$$

$$= x + x^2/2 + x^3/3 + \cdots. \qquad (9.42)$$

Since $-\ln (1 - x) = x + x^2/2 + x^3/3 \cdots$,

$$F(s) = -\ln (1 - s) = s + s^2/2 + s^3/3 + s^4/4 + \cdots \qquad (9.43)$$

One new A mutation per year maintains $1/n$ loci with n A's apiece, $\ln N$ segregating loci in all. As one new mutation per generation maintains $\ln N$ segregating loci in all, the average time elapsed before the descendants of a new A mutation either die out or spread to fixation is $\ln N$ generations.

When there are N individuals in the population, the genic variance at a locus with n A's is $n(N - n)/N^2$. As there is $1/n$ locus with n A's per new mutation, the total genic variance per mutation is

$$(1/N^2)[(N - 1) + (N - 2) + \cdots + 3 + 2 + 1]$$

$$= \tfrac{1}{2}N(N - 1)/N^2, \qquad (9.44)$$

which is very nearly $\tfrac{1}{2}$. As the genic variance introduced by a single new mutation is $(N - 1)/N^2$, $(1 - 2/N)$ as much as the total genic variance, it appears that here, too, genic variance would decline by $1 - 2/N$ per generation in the absence of mutation.

More generally, genic variance decays at a steady rate of $1 - v/N$ per generation when random changes in allele frequency are the only force for change, and one new, neutral mutation per year maintains roughly $(2/v) \ln N$ segregating loci (Fisher 1930a, 1930b; Haldane 1939; Harris 1963).

10 Effects of Chance on Allele Frequencies, II. Diffusion Equations

In population genetics, the "diffusion method" focuses on problems such as finding the probability $\phi(q, t)dq$ that, at time t, the frequency of an allele A lies between q and $q + dq$, or the probability $P(q)$ that if the allele A now has frequency q, it will attain the frequency 1. The diffusion method centers on reducing such problems to the solving of differential equations (preferably simple ones). Here, I will supply two examples.

To introduce the method, let us consider a locus with alleles A and a in a population of N mature haploids, where generations overlap. Let one individual be the young of another if the one inherits its allele at this locus directly from the other. Suppose, moreover, that every $1/N$ generations, an individual is chosen at random to produce one young, assumed to mature immediately, and that another individual (which might be the same one) is independently chosen to die and make room for the new adult. The number of A's can accordingly change by no more than 1, and their frequency by no more than $1/N$, between times t and $t + 1/N$ (here, time is measured in units of generations).

To be specific, suppose that, if there are j A's in the population at time t, the probability that the next individual to die is an A-bearer is j/N, as is the probability that the next individual to be born carries A. Then the probability $p(q, q + 1/N)$ that if A's frequency is $q = j/N$ at time t, it will be $q + 1/N$ at time $t + 1$, is the probability $1 - q$ an a dies, times the probability q that an A replaces her, or $q(1 - q)$. Thus

$$p(q, q + 1/N) = q(1 - q) = p(q, q - 1/N), \qquad (10.1)$$

$$p(q, q) = 1 - 2q(1 - q). \qquad (10.2)$$

Under these circumstances, the variance in change of A's frequency over $1/N$ time units is $V(q)/N$, where

$$V(q)/N = (1/N^2)[p(q, q + 1/N) + p(q, q - 1/N)]$$
$$= 2q(1 - q)/N^2. \qquad (10.3)$$

$V(q)$ is the variance in change of A's frequency over one generation, given that it is q to start with. Therefore,

$$V(q) = 2q(1 - q)/N, \qquad (10.4)$$

$$p(q, q + 1/N) = p(q, q - 1/N) = \tfrac{1}{2}NV(q). \qquad (10.5)$$

Suppose now that A affects the fitness of its bearers. If the probability that an A is chosen to die is $q[1 - k(1 - q)]$, while the corresponding probability for an a-bearer is $(1 - q)(1 + kq)$, then

$$p(q, q + 1/N) = q(1 - q)(1 + kq),$$
$$p(q, q - 1/N) = q(1 - q)[1 - k(1 - q)]. \qquad (10.6)$$

Here, the expected change in A's frequency between times t and $t + 1/N$, given that it is q at time t, is

$$M(q)/N = [p(q, q + 1/N) - p(q, q - 1/N)]/N$$
$$= kq(1 - q)/N. \qquad (10.7)$$

$M(q)$ is the mean change in A's frequency per generation, its mean change between times t and $t + 1$ if it is q at time t. The decreased mortality of A relative to a confers a selective advantage k on A over a. More generally, if we set

$$M(q) = [p(q, q + 1/N) - p(q, q - 1/N)], \qquad (10.8)$$

regardless of whether this change arises from selection, migration or mutation, then, in analogy with equation (10.6), we may set

$$p(q, q + 1/N) = \tfrac{1}{2}NV(q) + qM(q), \qquad (10.9)$$

$$p(q, q - 1/N) = \tfrac{1}{2}NV(q) - (1 - q)M(q). \qquad (10.10)$$

Let us imagine a large collection of islands, with exactly N haploids apiece. Let $\phi(q)/N$ be the proportion of these islands where A's frequency is exactly $q = j/N$. Suppose that each island exchanges a proportion m of its individuals as migrants each generation with a large mainland population where A's frequency is Q. Thus, for each island, a fraction m of the individuals is withdrawn each generation and replaced by individuals where A's frequency is Q, and

$$M(q) = m(Q - q). \qquad (10.11)$$

How effectively does the influx of migrants restrain random changes of A's frequency in the island populations?

The influx of migrants ensures that both A and a persist, at least intermittently, in each island population. We suppose accordingly that the system of islands is in equilibrium: that is to say, the proportion of islands in which A's frequency increases from q to $q + 1/N$ during time $1/N$, $[\phi(q)/N]p(q, q + 1/N)$, precisely balances the number $[\phi(q + 1/N)/N]p(q + 1/N, q)$ where A's frequency decreases from $q + 1/N$ to q. Using equations (10.9) and (10.10) to express the p's in terms of $M(q)$ and $V(q)$, our equation of balance becomes

$$\phi(q + 1/N)[\tfrac{1}{2}NV(q + 1/N) - (1 - q - 1/N)M(q + 1/N)]/N$$
$$= \phi(q)[\tfrac{1}{2}NV(q) + qM(q)]/N. \qquad (10.12)$$

Multiplying through by N and rearranging, equation (10.12) becomes

$$\tfrac{1}{2}[\phi(q + 1/N)V(q + 1/N) - \phi(q)V(q)]/(1/N)$$
$$= (1 - q - 1/N)M(q + 1/N)\phi(q + 1/N) + qM(q). \quad (10.13)$$

The upper line is very nearly $\tfrac{1}{2}(d/dq)V(q)\phi(q)$. If $(1 - q - 1/N)M(q + 1/N)\phi(q + 1/N)$ is very nearly $(1 - q)M(q)V(q)$, then equations (10.12) and (10.13) lead to the differential equations

$$\frac{1}{2}\frac{d}{dq}V(q)\phi(q) = M(q)\phi(q);$$

$$\frac{d}{dq}\ln V(q)\phi(q) = 2M(q)/V(q). \quad (10.14)$$

The solution to equation (10.14) is

$$\phi(q) = \frac{C}{V(q)}\exp 2\int_{q_0}^{q} M(x)dx/V(x), \quad (10.15)$$

where q_0 is a constant chosen for convenience, and C is chosen to ensure that ϕ is a genuine probability density, that is, that

$$\int_0^1 \phi(q)dq = 1. \quad (10.16)$$

Equation (10.15) was known to Wright (1931), even though he only discovered equation (10.14) later.

If we set $M(q) = m(Q - q)$, $V(q) = 2q(1 - q)/N$, then

$$2M(q)/V(q) = NmQ/q - Nm(1 - Q)/(1 - q), \quad (10.17)$$

$$\int_{q_0}^{q} 2M(x)dx/V(x) = Nm[Q \ln q + (1 - Q) \ln (1 - q)] + \ln C$$
$$(10.18)$$

$$\phi(q) = Cq^{NmQ-1}(1 - q)^{Nm(1-Q)-1}. \quad (10.19)$$

If NmQ and $Nm(1 - Q)$ both exceed 1, then $\phi(q)$ is maximum when $q = (NmQ - 1)/(Nm - 2)$, while if NmQ and $Nm(1 - Q)$ are both less than 1, then relatively few populations have substantial numbers of both A and a.

To assess the degree of variation between populations that genetic drift allows, let us compare the average genic variance within a population,

$$\bar{V}_g = \int_0^1 q(1 - q)\phi(q)dq, \tag{10.20}$$

with the variance in A's frequency over different populations,

$$S^2 = \int_0^1 (q - \bar{q})^2 \phi(q)dq, \quad \bar{q} = \int_0^1 q\phi(q)dq. \tag{10.21}$$

From a table of definite integrals, we use the fact that

$$\int_0^1 q^{a-1}(1 - q)^{b-1}dq = \frac{\Gamma(a)\Gamma(b)}{\Gamma(a + b)}, \tag{10.22}$$

where Γ is the "gamma function" (Artin 1964; Davis 1965), which has the useful property that $\Gamma(a + 1) = a\Gamma(a)$. Thus

$$C = \frac{\Gamma(Nm)}{\Gamma(NmQ)\Gamma[Nm(1 - Q)]},$$

$$\bar{V}_g = \frac{NmQ[Nm(1 - Q)]}{Nm(Nm + 1)} = \frac{NmQ(1 - Q)}{Nm + 1};$$

$$S^2 = \frac{NmQ(NmQ + 1)}{Nm(Nm + 1)} - Q^2 = \frac{Q(1 - Q)}{Nm + 1}, \tag{10.23}$$

$$S^2/\bar{V}_g = 1/Nm. \tag{10.24}$$

Variance of A's frequency among different populations exceeds the average genic variance per population only if less than one migrant from the mainland establishes itself per island population per generation (cf. equations 9 and 10 of Crow and Aoki 1982).

If $M(q) = kq(1 - q) + m(Q - q)$, then

$$\phi(q) = C'q^{NmQ-1}(1 - q)^{Nm(1 - Q)-1} \exp Nkq. \tag{10.25}$$

This probability density $\phi(q)$ for the values of A's frequency on different islands gives some idea of the relative strengths of the influences of genetic drift, selection, and migration from the mainland, on allele frequencies in these island populations. As we concluded in section 5, the selective differential k is of detectable importance relative to genetic drift when Nk exceeds 1, roughly speaking.

We can construct diffusion equations for a population where successive generations are distinct by imagining a "mimic" population with overlapping generations, where

$$p(q, q + 1/N) = \tfrac{1}{2}vq(1 - q) + qM(q);$$
$$p(q, q - 1/N) = \tfrac{1}{2}vq(1 - q) - (1 - q)M(q), \qquad (10.26)$$

where v is the variance in number of offspring per parent for the real population, and N is the number of reproductive adults in each generation. $V(q)$ is then $vq(1 - q)/N$. Strange as it may seem, the differential equations thus arrived at are accurate. Our fiction is equivalent to setting the "effective population size" N_e equal to N/v (Crow and Kimura 1970). Personally, I find the term "effective population size" can be misleading. The probability that a neutral mutation now represented by a single individual will spread to fixation is $1/N$ (the chances that her progeny will spread are exactly the same as those of each of her $N - 1$ fellows), not $1/N_e$. What changes is the "strength" of the genetic drift, and the speed with which allele frequencies respond to it.

In general, selection is of detectable importance relative to genetic drift if $2Nk/v$ exceeds 1; and variance among populations in A's frequency exceeds average genetic variance within populations if $2Nm/v$ exceeds 1.

Next, let us use diffusion methods to calculate the probability $P(q)$ that an allele whose frequency is currently $q = j/N$ will eventually attain frequency 1. To construct our differential equation, we decompose the probability $P(q)$ into the probability that in time $1/N$ A's frequency changes to $q + 1/N$, times the probability $P(q + 1/N)$ that it increases from $q + 1/N$ to 1; plus the probability that A's frequency stays at q during time $1/N$, times $P(q)$; plus the probability that A's frequency declines to $q - 1/N$ in time $1/N$, times the probability $P(q - 1/N)$ that it then increases from $q - 1/N$ to 1. Thus, using equations (10.9) and (10.10) for $p(q, q + 1/N)$ and $p(q, q - 1/N)$, we find that

$$P(q) = [\tfrac{1}{2}NV(q) + qM(q)]P(q + 1/N)$$
$$+ P(q) - [\tfrac{1}{2}NV(q) + qM(q)]P(q)$$
$$- [\tfrac{1}{2}NV(q) - (1 - q)M(q)]P(q)$$
$$+ [\tfrac{1}{2}NV(q) - (1 - q)M(q)]P(q - 1/N). \quad (10.27)$$

We may rewrite equation (10.27) as

$$0 = \tfrac{1}{2}NV(q)[P(q + 1/N) - P(q)]$$
$$- \tfrac{1}{2}NV(q)[P(q) - P(q - 1/N)]$$
$$+ qM(q)[P(q + 1/N) - P(q)]$$
$$+ (1 - q)M(q)[P(q) - P(q - 1/N)]. \quad (10.28)$$

Multiplying equation (10.28) by N and remembering, for example, that

$$N[P(q + 1/N) - P(q)] = dP/dq, \quad (10.29)$$

$$N^2\{[P(q + 1/N) - P(q)] - [P(q) - P(q - 1/N)]\} = d^2P/dq^2, \quad (10.30)$$

we find that

$$V(q)d^2P/dq^2 + 2M(q)dP/dq = 0, \frac{d}{dq}\ln\frac{dP}{dq} = -2M(q)/V(q), \quad (10.31)$$

$$\ln\frac{dP}{dq} = -\int_0^q 2M(x)dx/V(x) + \ln C, \quad (10.32)$$

$$P(q) - P(0) = C\int_0^q dy \exp -\int_0^y 2M(x)dx/V(x). \quad (10.33)$$

Since $P(0) = 0$, $P(1) = 1$, we find, as did Kimura (1962), that

$$P(q) = \frac{\displaystyle\int_0^q dy \exp -\int_0^y 2M(x)dx/V(x)}{\displaystyle\int_0^1 dy \exp -\int_0^y 2M(x)dx/V(x)}. \quad (10.34)$$

Let us first assume that $V(q) = 2q(1 - q)/N$, $M(q) = kq(1 - q)$. Then

$$\int_0^y 2M(x)dx/V(x) = Nky, \qquad (10.35)$$

$$P(q) = [1 - e^{-Nkq}]/[1 - e^{-Nk}]. \qquad (10.36)$$

If $\exp - Nk$ is far smaller than 1, while A is a new mutation represented by a single individual, so that $q = 1/N$, then $P(q) = k$. More generally, $P(1/N) = 2k/v$ if s is far less than 1, as we learned from the branching process method.

Now suppose that our population contains N diploids ($2N$ genes), and that A is a recessive allele with selective advantage $k = Kq$. Then $V(q) = vq(1 - q)/2N$, and

$$\int_0^y 2M(x)dx/V(x) = 2Nky^2/v. \qquad (10.37)$$

If $\exp - 2NK/v$ is far smaller than 1, we may set

$$\int_0^1 dy \exp - 2NKy^2/v$$

$$\approx \sqrt{\pi v/2NK}\left[\sqrt{4NK/2\pi v}\int_0^\infty dy \exp - 2NKy^2/v\right]. \qquad (10.38)$$

The expression in brackets is half the integral of a normal distribution, or $1/2$ (since $\exp - 2NK/v$ is so very small compared to 1, extending the integral from 1 to ∞ makes almost no difference). Since

$$\int_0^{1/2N} \exp - 2NKy^2/v \approx 1/2N, \qquad (10.39)$$

we find that

$$P(1/2N) = (1/2N)/[\tfrac{1}{2}\sqrt{\pi v/2NK} = \sqrt{2K/N\pi v} \qquad (10.40)$$

(Kimura 1964; Crow and Kimura 1970). The probability of fixation of a new recessive mutation is roughly the geometric mean of the probability $\frac{1}{2}N$ that a new neutral mutation will spread to fixation, and the probability K/v that a new mutation of constant selective advantage $\frac{1}{2}K$ will do so.

At the moment, diffusion equations are our most versatile tools for discovering how genetic drift influences the frequencies of alleles undergoing various types of selection. Unfortunately, when Haldane was writing, Fisher (1922, 1930a), who had introduced the method, had expressed his equations in terms of the variable θ, the angle whose cosine is $1 - 2q$. Variance in change of θ per generation is a constant, independent of q, but the expression for the mean change in θ proved to be unexpectedly complicated (Fisher 1930a). Wright (1931) used integral equations to derive his earlier results. Wright (1945a) did reintroduce diffusion equations, following a lead from the Russian mathematician Kolmogorov, but Kimura (1964) was the first to show how useful the method could be.

11 Conflicts Between Selection at Different Levels

The latent or overt conflict between an individual's advantage and the good of the group of species to which he belongs is as old as humanity. The "tragedy of the commons" (Hardin 1968), whereby the benefit immediately accruing to each individual farmer from stocking the common pasture with one more animal than he should, ruins the pasture through overstocking, is but one manifestation of this conflict. More generally, inappropriate forms of competition may benefit an individual or a family, even though they are ruinous to society. The problem of ethical, political, and economic theory, from the days of Plato's *Republic* onward, has been to discover either the coincidence of enlightened self-interest with the good of society, or the mode of organizing society that most nearly aligns individual advantage with the good of the community.

The conflict between individual advantage and the good of society finds parallels all through biology. It benefits a worker wasp to lay eggs (if she can "get away with it") rather than to help raise her queen's young, even though this behavior is hurtful to the colony as a whole. In many species, the survival or growth of the population depends either on the number of females raised or on the amount of effort expended on each one raised. These populations need only enough males to assure the competent fertilization of the females. Yet, because a successful sperm contributes just as much to the genetics of future generations as a successful egg, populations spend half their reproductive effort raising males.

A. The conflict between individual advantage and the good of the group or species

Unlike such problems as the balance between mutation and stabilizing selection on a quantitative characteristic, the conflict between individual and group selection has only recently become part of the repertoire of theoretical population genetics (Slatkin and Wade

1978; Aoki 1982; Crow and Aoki 1982, 1984; Kimura 1983b). Haldane was not the only one of the earlier population geneticists to concern himself with the conflict between selection in, and of, populations: so did Wright (1945b), Fisher (1930a, 1958a), and Lewontin (1962). Analysis of this conflict, however, involves combining Wright's concept of genetic drift and Fisher's fundamental theorem of natural selection. The tools that would solve the problem were available in 1931, but it was years before they were used. Wright (1945b) saw that equation (10.19) offered the key to this problem, but he did not use the key to turn the lock. Did he, like Haldane, dismiss Fisher's fundamental theorem from his mind? Fisher (1958a, p. 49) was mesmerized by equations analogous to (4a.15) into thinking that relative fitness within populations was the only selective influence on allele frequencies. Was Fisher blinded as well by contempt for genetic drift? For, as we shall see, Haldane (and Wright) were absolutely correct to see genetic drift as the source of variation among populations on which selection of populations can act.

In the end, the first theoretical attempts to discern what circumstances would allow selection of populations to override selection within populations came from outside population genetics (Maynard Smith 1964; Levins 1970; Leigh 1971; Wilson 1977; Levitt 1978).

Haldane (1932, pp. 119–122) modeled the conflict between individual advantage and the good of the population by considering a single locus, with an "altruist" allele A and a "selfish" allele a, in a species divided into isolated subpopulations. He assumed that individuals of altruist phenotype had fitness $1 - k$ relative to selfish ones, but that population numbers multiplied by $1 + KQ$ per generation, where Q is the proportion of individuals in the population of altruist phenotype. Here, k measures the cost to an individual of possessing this altruistic feature, and KQ represents the net benefit to the population from a proportion Q of its individuals possessing the altruistic feature. He learned that A's frequency must necessarily decrease, even though A's numbers would increase as long as $(1 - k)(1 + KQ)$ exceeds 1, that is, as long as Q exceeds $k/K(1 - k)$.

If we suppose that our species is haploid, so that the proportion $Q(n)$ of altruists in generation n is simply the proportion of A's in the population, A's numbers increase if its frequency Q exceeds $k/K(1 - k)$, but its frequency declines according to the law

$$Q(n + 1) = \frac{Q(n)(1 - k)}{1 - kQ(n)};$$

$$Q(n + 1) - Q(n) = -\frac{kQ(n)[1 - Q(n)]}{1 - kQ(n)}. \qquad (11.1)$$

Eventually, A's frequency declines to the point where A's numbers do too.

For Williams (1966), this settled the matter. He cited Haldane as one of his authorities for dismissing selection between populations as a force not worth worrying about. Haldane, however, held open the possibility that the altruistic allele A could survive if each population was small enough that genetic drift would sometimes increase A's frequency (1932, pp. 121), or if the altruism of A-bearers was directed toward their relatives (p. 119). We will soon see how related these two suggestions are.

To learn when genetic drift will allow selection of populations to override selection within populations against altruistic alleles, imagine a species divided into T populations of N adult haploids apiece. Let successive generations be distinct. Suppose that selection acts on a quantitative characteristic, whose measure is x, governed by many loci. Let individuals with higher x leave relatively more offspring in the next generation of their population, and let populations with lower average values of x be less likely to go extinct. To be specific, let an individual with value x of this characteristic leave, on the average, $\exp k(x - y)$ times as many offspring as an individual with value y, and let the chance that a population suddenly disappears between generations n and $n + 1$ be $E \exp K(\bar{x} - \bar{\bar{x}})$, where $\bar{x}$ and $\bar{\bar{x}}$ are the average values of the characteristic in the population and in the species as a whole, and E is a constant.

Assume that in each population, values of the characteristic are normally distributed, as is usual for characteristics determined by many genes (Falconer 1981). Let the variance in these values be the same amount V in every population. Finally, let us suppose that the value of the characteristic is precisely determined by the genes that affect it.

Under these circumstances, the change in a population's mean relative fitness per generation, $k[\bar{x}(n + 1) - \bar{x}(n)]$, is k^2V, the genic variance in relative fitness, as equation (4a.20) suggests. Thus

$$\bar{x}(n + 1) - \bar{x}(n) = kV. \qquad (11.2)$$

Moreover, in *this* model, selection does not change the variance. To show this, let the probability that, within a given population, an individual's value lies between x and $x + dx$ be

$$C dx \exp - (x^2 - 2x\bar{x})/2V, \qquad (11.3)$$

where $C = 1/[\sqrt{2\pi V} \exp \bar{x}^2/2V]$. Then the corresponding probability density for generation $n + 1$ is

$$C' dx \exp [kx - (x^2 - 2x\bar{x})/2V]$$
$$= C' dx \exp - [x^2 - 2x(\bar{x} + kV)/V]. \qquad (11.4)$$

This latter is indeed a normal distribution with mean $\bar{x} + kV$ and variance V, as was promised.

If all surviving populations send out equal numbers of migrants each generation, and have equal chances of contributing colonists to establish new populations, then a population's fitness (its total contribution of genes to future populations) is proportional to its lifetime. The average lifetime of a population with mean value $\bar{x}$ is $1/[E \exp K(\bar{\bar{x}} - \bar{x})]$ generations. The variance in relative fitnesses of populations is thus K^2 times the variance S^2 in their mean values of x. In analogy with selection between individuals, selection between populations changes the mean value $\bar{\bar{x}}$ for the species as a whole by an amount KS^2 per population lifetime.

The effect on our quantitative characteristic of selection of populations relative to selection within populations is the change per individual generation from selection of populations, KS^2/L, where L is average population lifetime, relative to change per generation from selection within populations, kV. This ratio is KS^2/LkV. The crucial unknown in this relationship is S^2/V.

This result can be interpreted in terms of the relatedness of members of the same population (Crow and Aoki 1982). The intraclass correlation coefficient (Fisher 1958b, pp. 211–213) between members of the same population is

$$r = S^2/(V + S^2). \qquad (11.5)$$

This coefficient r measures the relatedness, the genetic correlation, between two members of the same population, relative to two members chosen at random from the species as a whole. A unit decrease in x multiplies the fitness of an individual by $1 - k$, while a unit decrease in a population's average, $\bar{x}$, multiplies the number of offspring populations per parent population per individual generation by $1 + K/L$. Thus k is the cost to an individual of a unit increase in x, while $b = k + K/L$ is the benefit to each member of the group from a unit increase in $\bar{x}$ (K/L is the excess $b - k$ of benefit over cost). The benefit to the group outweighs the cost to its individuals if

$$kV < KS^2/L = (b - k)S^2, \tag{11.6}$$

that is to say, if

$$kV < (b + K)S^2;$$
$$k < bS^2/(V + S^2) = br. \tag{11.7}$$

This statement embodies the general truth that, if the ratio of the reproductive cost k of helping a relative to the reproductive benefit b thereby conferred on that relative is less than the coefficient r of relationship between helper and beneficiary, it pays to help (Hamilton 1964). In particular, if br exceeds k, this help will, on balance, spread the benefactor's genes, for br represents the spread of the benefactor's genes from the help given the relative. Equation (11.7) is the foundation of the theory of kin selection (West Eberhard 1975).

Let us first apply equation (11.6) to "trait-group selection" at Haldane's altruistic locus, with its alleles A and a. Suppose, with Wilson (1977), that the adults of our species divide into "trait groups" of N individuals apiece, and that an individual's reproduction is governed by interactions within her trait group. Let the young join a common pool, from which the next generation's trait groups are assembled at random. How does the good of a trait group weigh against the advantage of one of its individuals? The variance S^2 among trait groups in A's frequency is the variance in means of samples of N from a distribution with variance V^*, or V^*/N, variance V within populations is $(N - 1)V^*/N$, and group lifetime is one generation. Thus trait group selection balances individual selection when $K/(N - 1) = k$. Trait group selection does not even require a species to be divided into discrete groups. It only requires than an individual's reproduction be governed by her interactions with a restricted number of neighbors (Wilson 1980, p. 38).

If K is positive and k is 0, so that A-bearers benefit all members of their trait group equally, regardless of genotype, at no cost to themselves, then trait group selection favors A. In herbivores, such "costless benefits" can be achieved by adjusting the composition of placement of feces so as best to fertilize their food plants. Earthworms may confer such costless benefits by adjusting their behavior so as to improve the soil and promote plant growth, thereby increasing the quantity of leaf litter fall (and therefore the amount of earthworm food) for all genotypes alike. In a uniform, random-mating population, an allele derives no advantage by benefiting individuals of all trait groups equally. Trait group selection may therefore play a role in the evolution of the mutualistic inter-dependences so characteristic of ecological communities (Wilson 1980).

Next, let us suppose that our species is divided into sub-populations that are distinct and long-lasting. Let us consider a quantitative characteristic governed by so many loci that selection at any one locus is negligible compared to genetic drift. Let us also suppose that variance in each population is stabilized by the exchange of migrants at random among populations. Even though allele frequencies at any one locus vary greatly from population to population, there are so many loci that the total variance is the same in all populations. In this case, migrants will communicate the effects of selection within and between populations to all parts of the species. Although we derived equation (10.24) assuming that the average Q of A's frequency among the migrants to a set of island populations was its frequency in a large mainland population, we could just as well assume, with Wright (1945b), that there is no mainland, that the islands exchange migrants with each other, and that Q is the average of A's frequency for the species as a whole. If populations live long enough that the buildup of genic variance in a newly established population takes only a small part of its lifetime, then equation (10.24), an immediate consequence of equation (10.19), which was first derived by Wright (1931), shows that

$$S^2/V = v/2Nm, \qquad (11.8)$$

where v is the variance in the number of offspring per parent, and m is the proportion of migrants exchanged at random through the species. If migration is the factor governing the value of S^2/V, and if individual and interpopulation selection are of equal strength, so that $k = K$, then selection of populations can override selection within populations if

$$v/2NmL < 1. \qquad (11.9)$$

When $v = 1$, as in the models of Fisher and Wright, selection of populations overrides selection within populations of equal intensity only if less than one migrant is exchanged per two populations per population lifetime. This conclusion has been derived much more rigorously by Aoki (1982).

Let us now follow Leigh (1983, 1986), and consider more generally how the variance S^2 among populations is related to within-population variance V. First, consider how S^2 changes between generations n and $n + 1$. The means of samples of N from a normal distribution with variance V has variance V/N. Therefore, the vagaries of sampling N individuals of one generation from the young of the previous generation adds an amount V/N each generation to S^2. Let exchange of migrants replace a fraction m of each population's adults with individuals of average value $\bar{\bar{x}}$, thus reducing S^2 by a factor $(1 - m)^2$, or roughly $1 - 2m$. Next, let each extinct population be replaced immediately by a new population where colonists from M randomly chosen parent populations are equally represented. Then, between generations n and $n + 1$, a fraction $E = 1/L$ of the populations in this species are replaced by populations the variance of whose means is S^2/M. If each new population represents two parent populations, one contributing a fraction p of the colonists, the variance among these new populations is $S^2[p^2 + (1 - p)^2]$. In this case, we may set $M = 1/[p^2 + (1 - p)^2]$. Finally, just as the chance of who happens to reproduce reduces the variance of a haploid population with overlappng generations by the factor $1 - 2/N$ per generation, the chances of which populations "reproduce" reduces S^2 by a factor $1 - 2/MLT$ per individual generation. Then

$$S^2(n + 1) \approx S^2(n)(1 - 2m)\left(1 - \frac{2}{MLT}\right)$$

$$+ S^2(n)\left(\frac{1}{M} - 1\right)\frac{1}{L} + \frac{V}{N}. \tag{11.10}$$

If variance among populations has attained its equilibrium, so that $S^2(n + 1) = S^2(n) = S^2$, and if $4m/MLT$ is negligibly small, then

$$S^2\left[2m + \frac{2}{MLT} + \left(1 - \frac{1}{M}\right)\frac{1}{L}\right] \approx V/N. \tag{11.11}$$

S^2/L exceeds V, and selection of populations can override an equally intense selection within populations, if

$$2mNL + \frac{2N}{MT} + N\left(1 - \frac{1}{M}\right) < 1. \qquad (11.12)$$

If migration is the only influence reducing S^2, then, as we have already remarked, S^2/L exceeds V if less than one migrant is exchanged per two populations per population lifetime. If $m = 0$ and T, the total number of populations, is very large, then S^2/L exceeds V only if $N(1 - 1/M)$ is less than 1. If each population is founded by two very unequally represented parent populations, the rarer one very rare indeed, with frequency p much less than 1, then $N[1 - p^2 - (1 - p)^2]$, which is very nearly $2Np$, must be less than 1 for selection of populations to override an equal individual selection. This condition implies that each population can have only one parent population for selection of populations to prevail. Finally, if $M = 1$, $m = 0$, then S^2/L exceeds V only if T exceeds $2N$, that is, if there are more than twice as many populations as individuals per population.

Parasites that evolve to exploit a host organism more carefully, rather than depending primarily on the ability to keep dispersing from one individual to another, often evolve characteristics that allow group selection to override selection within populations. Although intracellular organelles possess their own genes and replicate autonomously, the inheritance of organelles only through the mother and the impossibility of exchanging migrants even among cells (Eberhard 1980) ensure that they evolve primarily by selection of organelle populations. Their mode of "reproduction" and "dispersal" ensures that their fate is identified nearly completely with that of their hosts. Although some organelles began as parasites, they have become partners in a mutualism so absolute that it is hard to believe their ancestors were independent organisms (Margulis 1981).

It appears, however, that selection of populations can override selection within populations only under most unusual circumstances. Lewontin (1962) demonstrated an example of the effectiveness of group selection, but the conditions that allowed interdeme selection to be effective seemed remarkably restrictive. As a result, most biologists have focused exclusively on individual selection.

Even if it cannot override individual selection, selection between species may be an important process. Those species where individual advantage conflicts most with the good of the population presumably go extinct most quickly (Leigh 1977). Does this process reconcile individual advantage with the good of the species? Is there evidence of such a reconciliation? These questions are not yet answered.

B. *The conflict between selection on genes and selection on individuals*

Just as selection sometimes favors individuals with characteristics detrimental to their populations, selection sometimes favors genes with characteristics detrimental to the individuals that carry them. Consider, for example, a locus with alleles A and a in a population where successive generations are distinct. Let the relative fitnesses of AA, Aa, and aa be $1 - K$, 1, and 1, respectively. Assume, however, that A-genes somehow bias meiosis in their own favor, so that a fraction $\frac{1}{2}(1 + D)$ of the gametes produced by Aa carry A. Thus, if A's frequency among the adults of generation n is $q(n)$,

$$q(n + 1) = \frac{(1 - K)q^2(n) + q(n)[1 - q(n)](1 + D)}{1 - Kq^2(n)},$$

(11.13)

$$q(n + 1) - q(n) = \frac{q(n)[1 - q(n)][D - Kq(n)]}{1 - Kq^2(n)}.$$

(11.14)

When rare, the "segregation-distorter allele" A spreads even though it confers a phenotypic defect on some of its bearers. Indeed, if D exceeds the selective disadvantage of AA phenotypes, A will spread through the population, to the detriment of every individual in it. The spread of such distorter alleles can cause the extinction of whole populations (Werren, Nur, and Wu 1988).

In theory, an allele could cause an apparent bias in segregation ratios of males by causing the sperm carrying it to race faster to the egg. As Haldane (1932, p. 67) remarked, this possibility is excluded, at least in *Drosophila*, by investing the control of a sperm's behavior in its father's genotype, not its own. In many species, a male makes so many sperm that half of them can be destroyed without materially affecting his ability to fertilize females. In these species, an allele can create an apparent bias in the meiosis of heterozygous males by destroying sperm that carry the alternative allele. Many of the segregation distorters that are known (*t*-haplotypes in mice, *SD* in *Drosophila*) appear to act in this manner (Brown *et al.* 1989; Crow 1988).

Let us return to our distorter A, whose homozygotes are defective. An allele on another chromosome derives no benefit from A's distorted segregation ratios, because alleles on different chromosomes assort independently. A mutant on a chromosome different from A's can spread by suppressing A's distortion effect, by restoring, as it were, the honesty of meiosis, for by doing so, the mutant spares some of its bearers AA's phenotypic defect, which segregation distortion would otherwise have spread to them (Prout, Bundgaard, and Bryant 1973). As mutants restoring the honesty of meiosis are favored at any locus not linked to A, the honesty of meiosis can be said to represent the common interest of the genome as a whole. In animals, this selection appears to have been quite effective (Leigh 1987). The control of sperm behavior by the paternal genotype is one consequence of this selection (Crow 1979). The honesty of meiosis ensures that an allele spreads only if it benefits the individuals carrying it: in effect, honest meiosis aligns the advantage of each allele with the good of the individuals carrying it. The question of how selection at different levels is reconciled never occurs in Haldane's book, but it may be a fundamental process in evolution.

Nowadays, transposable elements have provided new examples of genes that spread without regard to whether they benefit their carriers (Engels 1986; Charlesworth 1988; Werren, Nur, and Wu 1988). These are much less obviously subordinate to the good of their carriers than genes on chromosomes, and they have most decidedly attracted the attention of theoretical population geneticists.

In animals, meiosis is honest, and gametes are the instruments of their parents. In plants, only the first is true. Pollen can compete with both other pollen from the same plant, and pollen from other plants. The structure of pollen can also evolve to improve the ease of transmission. The common interest of the genome as a whole presumably favors parental control of gametic behavior, as in animals. Nevertheless, as Haldane (1932, p. 67) remarks, the behavior of a pollen grain is determined, at least to some extent, by its own genes. What role selection on pollen plays in how plants evolve still seems far from clear (Snow and Mazer 1988).

12 MIGRATION AND SPECIATION

Let us consider, with Haldane (1932, pp. 122–123), a locus with alleles A and a in an island population that receives migrants from the mainland. What happens when A-bearers are more fit on the island, while all the immigrants carry a? Suppose first that the relative fitnesses of AA, Aa, and aa on the island are 1, $1 - k$, and $(1 - k)^2$, respectively, and that, each generation, a fraction m of the reproductive adults on the island are replaced by aa from the mainland. Then

$$q(n + 1) =$$

$$\frac{\{q^2(n) + (1 - k)q(n)[1 - q(n)]\}(1 - m)}{q^2(n) + 2(1 - k)q(n)[1 - q(n)] + (1 - k)^2[1 - q(n)]^2}$$

$$= \frac{q(n)(1 - m)}{1 - k[1 - q(n)]}. \quad (12.1)$$

$$q(n + 1) - q(n) = \frac{kq(n)[1 - q(n)] - mq(n)}{1 - k[1 - q(n)]}$$

$$= \frac{q(n)\{k[1 - q(n)] - m\}}{1 - k[1 - q(n)]} \quad (12.2)$$

Despite their selective advantage, A's cannot maintain themselves on the island unless k exceeds m. If k does exceed m, then $q(n)$ increases until $k[1 - q(n)]$ attains the value m, that is, until $1 - q(n)$ attains the value m/k. At equilibrium,

$$1 - q = m/k; \quad (12.3)$$

a's frequency $1 - q$ is the smaller the larger k is relative to m.

Now suppose that AA is recessive, and that

$$q(n + 1) = \frac{\{q^2(n) + (1 - K)q(n)[1 - q(n)]\}(1 - m)}{1 - K[1 - q^2(n)]}$$

$$= \frac{\{q(n) - Kq(n)[1 - q(n)]\}(1 - m)}{1 - K[1 - q^2(n)]}, \quad (12.4)$$

$$q(n + 1) - q(n) = \frac{q(n)\{Kq(n)[1 - q(n)] - m + mK[1 - q(n)]\}}{1 - K[1 - q^2(n)]}$$

$$= \frac{q(n)B(q)}{1 - K[1 - q^2(n)]}. \quad (12.5)$$

A's frequency q increases only if $B(q)$ is positive, that is, only if

$Kq(1 - q)$ exceeds $m[1 - K(1 - q)]$. A is maximum when $dB/dq = 0$, which holds when

$$K(1 - 2q) - mK = 0; q = (1 - m)/2. \tag{12.6}$$

If B is not positive when $q = (1 - m)/2$, it will never be. To see this, set $q = (1 - m + x)/2$. Then

$$B = \tfrac{1}{4}K[1 - (m - x)^2] - m + \tfrac{1}{2}(1 + m - x)mK$$
$$= \tfrac{1}{4}K(1 + m)^2 - m - \tfrac{1}{4}Kx^2. \tag{12.7}$$

Thus, the larger x, the smaller B: B is maximum when $x = 0$. Equation (12.7) allows us to draw two conclusions.

First, if $K(1 + m)^2$ is less than $4m$, A must necessarily disappear from the island.

Second, if $K(1 + m)^2$ exceeds $4m$, then A increases if x^2 is less than $(1 + m)^2 - 4m/K$, that is, if q is within $\tfrac{1}{2}[(1 + m)^2 - 4m/K]^{\frac{1}{2}}$ of the value $(1 - m)/2$. If this is so, then A's frequency increases to the equilibrium value $\tfrac{1}{2}(1 - m) + \tfrac{1}{2}[(1 + m)^2 - 4m/K]^{\frac{1}{2}}$. Likewise, if q exceeds this equilibrium value, it declines to it. On the other hand, if q is less than $\tfrac{1}{2}(1 - m) - \tfrac{1}{2}[(1 + m)^2 - 4m/K]^{\frac{1}{2}}$, that is, if $B(q)$ is negative because q is too low, then the A's inexorably disappear from the island.

A's presence on the island thus depends not only on the current rate of influx of migrants from the mainland, but also on the previous history of both the island and mainland populations. If A is present now on the island, either A was once favored on the mainland, or the island was once completely isolated from the mainland, or the island's population was once small enough, and migrants rare enough, that A could have spread by chance.

One can expand on this result in two ways, which are most easily illustrated by assuming that our species is haploid. First, suppose that after a balance has been struck between selection for A on the island and immigration of a from the mainland, a new mutant B appears in the island population, at a locus on the same chromosome as A. It turns out that A's presence reduces the selective differential needed to maintain B in the population (Slatkin 1981).

To see this, consider selection jointly on two loci, one with alleles A, a, and one with alleles B, b, as in section 7. Let x_{AB}, x_{Ab}, x_{aB}, and x_{ab} be the frequencies of AB, Ab, aB, and ab among the island's juveniles of generation n; let the proportions of these genotypes that grow to maturity be 1, $1 - K$, $1 - k$, and $(1 - K)(1 - k)$, respectively; let the survivors mate at random; and let a proportion m of the newborns of each genotype be exchanged for ab's from the mainland. What are the frequencies of these genotypes among the island's juveniles of the generation following?

The frequencies of these genotypes among the adults of generation n will be

$$x''_{AB} = x_{AB}/\bar{W}, x''_{Ab} = x_{Ab}(1 - K)/\bar{W}, x''_{aB} = x_{aB}(1 - k)/\bar{W},$$
$$x''_{ab} = x_{ab}(1 - k)(1 - K)/\bar{W}. \tag{12.8}$$

$\bar{W}$ is the mean fitness of these adults:

$$\bar{W} = x_{AB} + x_{Ab}(1 - K) + x_{aB}(1 - k) + x_{ab}(1 - K)(1 - k)$$
$$= 1 - K(1 - q) - k(1 - p) + kKx_{ab}. \tag{12.9}$$

Here, $p = x_{AB} + x_{Ab}$ is A's frequency among the juveniles of generation n, and $q = x_{AB} + x_{aB}$ is B's. Let r be the frequency of recombination between the two loci. Using table 7.1, and letting $D = x_{AB}x_{ab} - x_{Ab}x_{aB}$, we may express the frequencies of these genotypes among the newborns of generation $n + 1$ as

$$x'_{AB} = [x_{AB} - r(1 - K)(1 - k)D/\bar{W}]/\bar{W}, \tag{12.10}$$

$$x'_{Ab} = [x_{Ab}(1 - K) + r(1 - K)(1 - k)D/\bar{W}]/\bar{W}, \tag{12.11}$$

$$x'_{aB} = [x_{aB}(1 - k) + r(1 - K)(1 - k)D/\bar{W}]/\bar{W}. \tag{12.12}$$

Now set $p' = x'_{AB} + x'_{Ab}, q' = x'_{AB} + x'_{aB}, x_{Ab} = p(1 - q) - D$, and $x_{aB} = (1 - p)q - D$, and express mean fitness as

$$\bar{W} = [1 - K(1 - q)][1 - k(1 - p)] + kKD. \tag{12.13}$$

Then, if we neglect kK compared to 1, we find that

$$p' = [p - Kx_{Ab}]/\bar{W} = p[1 - K(1 - q)] + KD/\bar{W}$$
$$\approx p/[1 - k(1 - p)] + KD/\bar{W}, \tag{12.14}$$

$$q' = [q - kx_{aB}]/\bar{W} \approx q/[1 - K(1 - q)] + kD/\bar{W}. \tag{12.15}$$

If D is positive, that is, if B's presence is correlated with A's, the spread of each allele is enhanced by the presence of the other, but the allele of less effect receives the greater benefit from the association. The influx of migrants from the mainland tends to keep D positive, as we shall now show.

After the exchange of migrants, the frequencies of AB, A, and B become

$$x_{AB}(n + 1) = x'_{AB}(1 - m);$$
$$p(n + 1) = p'(1 - m); q(n + 1) = q'(1 - m), \quad (12.16)$$

respectively. The linkage disequilibrium in generation $n + 1$ is

$$D(n + 1) = x_{AB}(n + 1) - p(n + 1)q(n + 1)$$
$$= D(n)(1 - K)(1 - k)(1 - r)(1 - m)/\bar{W}^2$$
$$+ m(1 - m)p'q'. \quad (12.17)$$

Now suppose that B is very rare, so that $\bar{W} = (1 - K)$ $[1 - k(1 - p)]$. Suppose also that A's frequency is in balance between selection and migration, so that $p(1 - m)/$ $[1 - k(1 - p)] = p$. Then

$$1 - k(1 - p) = 1 - m; \bar{W} = (1 - K)(1 - m), \quad (12.18)$$

$$D(n + 1) = D(n)(1 - k)(1 - r)/\bar{W} + m(1 - m)p'q'. \quad (12.19)$$

When $D(n + 1) = D(n) = D$,

$$D[(1 - K)(1 - m) - (1 - k)(1 - r)]$$
$$= m(1 - m)p'q'\bar{W} \approx mpq. \quad (12.20)$$

Then, if B is rare and mK negligibly small, equations (12.15) and (12.16) suggest that

$$q(n + 1) \approx q(n)(1 - m)\left[1 + K + \frac{kmp}{k(1 - r) + r - K - m}\right]$$

$$\approx q(n)(1 - m)\left[1 + K + \frac{m(k - m)}{k(1 - r) + r - K - m}\right]. \quad (12.21)$$

If B is closely linked to A, that is, if r is less than k, and if k is several times larger than m, A's presence halves, at least, the selective advantage required to maintain B on the island. Thus, once A establishes itself securely in the face of migration, it becomes a "magnet" for other genetic differences at nearby loci, as Slatkin (1975) found in a different context.

   Now let us suppose, with Slatkin (1981), that A has a pleiotropic effect (a "side-effect") that diminishes the fertility of matings with a. Perhaps the genetic difference impairs the effectiveness of fertilization, or renders courtship time consuming but fruitless. If fertility of such "hybrid matings" is low enough, A cannot spread when rare, for when rare it mates mostly with a, suffering the disadvantages of hybridizing, while the a's, being common, mate mostly with each other. On the other hand, if the A's are securely established on their island, and if their selective advantage k over a-bearers is several times the rate m of influx of migrants, the penalty of hybridizing falls mainly on the rare immigrating a's, and the reproductive incompatibility increases A's frequency. Just as if A were recessive, the hybrid disadvantage renders A's presence dependent upon the previous history of the island.

Suppose, to be specific, that on the island, a's survive to maturity $1 - k$ times as well as A's, while matings between A and a bear $1 - t$ times as many offspring apiece as do "homozygous" matings. If A's frequency among the juveniles of generation n is q, its frequency at mating time will be $q'' = q/[1 - k(1 - q)]$, and its frequency among the resulting newborns will be

$$q' = \frac{q''[q'' + (1 - t)(1 - q'')]}{1 - 2tq''(1 - q'')}$$

$$= \frac{q^2 + q(1 - q)(1 - t)(1 - k)}{[1 - k(1 - q)]^2 - 2tq(1 - q)(1 - k)}. \tag{12.22}$$

Setting $q(n + 1)$, A's frequency among the juveniles of generation $n + 1$, equal to $q'(1 - m)$, and neglecting terms in k^2, tk, tm, km, etc., we find that

$$q(n + 1) - q(n)$$
$$= q(n)[1 - q(n)]k + t[2q(n) - 1] - mq(n). \tag{12.23}$$

If $k - t$ exceeds m, A can spread even when rare. More generally, A's frequency is in equilibrium if

$$B(q) = [k + t(2q - 1)](1 - q) - m = 0. \tag{12.24}$$

$B(q)$ is maximized when $dB/dq = 0$, that is, when $q = (3t - k)/4t$. Setting $q = x + (3t - k)/4t$, we find that

$$B(q) = (k + t)^2/8t - m - 2tx^2. \tag{12.25}$$

A cannot possibly be maintained in the population unless $(k + t)^2$ exceeds $8tm$. If this is true, and if A's frequency exceeds $(3t - k)/4t - c$, where

$$c = (1/4t)[k + t)^2 - 8tm]^{\frac{1}{2}}, \tag{12.26}$$

then A's frequency will approach $(3t - k)/4t + c$. Otherwise, A's frequency declines to 0.

Consider now a new allele B, favored on the island but not on the mainland, where matings between B and b, like those between A and A, are less fertile than "homozygous" matings. If B is sufficiently common to begin with, A's presence makes it easier for B to maintain itself in the presence of immigrating b's, because B will tend to occur primarily in A-bearers (Slatkin 1981). This association of B with A increases the reproductive isolation between island and mainland forms. Slatkin's model shows how speciation might arise as an accidental consequence of genetic divergence.

The problem of Haldane with which this section started is also a precursor of the many theoretical studies of clines—environmental gradients along which the selective advantage of A over a gradually reverses, with A strongly favored at one end and a at the other. Whether clines are zones of incipient fission, where one species is gradually becoming two, has long been a question of interest (Fisher 1930a; Endler 1977). Perhaps as a result, clines, like mutation-selection balances in polygenic traits, have become a "standard problem" of population genetics (Haldane 1948; Fisher 1950; Slatkin 1973, 1975; Nagylaki 1975; Felsenstein 1975; May, Endler, and McMurtrie 1975; Barton 1986a).

13 WRIGHT'S THEORY

Sewall Wright was very sensitive to the complexities of gene interaction. Wright (1934) foresaw that genes programmed chemical reactions, presumably by programming the enzymes that catalyzed them. Since each gene alters the chemical background for all the others, how could one predict their interactions?

Therefore, an allele's contribution to the fitness of one of its bearers depends on what other genes that bearer carries. Haldane (1932, Table V, p. 56) gives examples of this dependence. Wright argued that, to facilitate adaptive evolution, circumstances must allow selection to test gene combinations, not merely the average merits of individual genes. Wright (1931, p. 145) thought that adaptation would be most rapid and effective if a series of asexual generations alternated with an occasional sexual generation: sexual reproduction generates new gene combinations, which cloning reproduces for testing by selection. Wright (1931, p. 145) remarked that the alternation of sexual with asexual reproduction "has been a favorite [method] of the plant breeder and is perhaps the most successful yet devised for the human control of evolution in those cases where it can be applied at all."

What of populations with no "alternation of generations"? In very small populations, genetic drift overpowers selection. Wright (1931) thought that, in very large populations, selection at each locus would spread the allele that was best "on the average," eliminating the others, leaving little scope for forming new gene combinations. Later, Wright (1932) emphasized that selection would strand large populations on the nearest "adaptive peak," from which they could never progress to higher peaks. For example, a large haploid population, now of genotype ab, could not replace it with the more fit genotype AB if aB and Ab were both less fit than ab. In Wright's language, a large, randomly mating population could not cross the "adaptive valley" from the peak ab to the higher peak AB.

Wright argued that the situation is far more favorable for species divided into many small populations, each population isolated enough that frequencies of weakly selected alleles vary greatly from population to population, but with sufficient exchange of migrants to prevent the fitness of any population from deteriorating excessively. To generate as many new gene combinations as possible, populations must be small enough to allow genetic drift to "drive" allele frequencies across shallow "adaptive valleys". Nevertheless, populations should be large enough that, if a new gene combination proves particularly advantageous, selection will spread it through the population. The resulting overflow of migrants will then spread it to the rest of the species (Wright 1931, p. 151).

Wright's 1931 paper was the only statement of Wright's view on evolution available to Haldane when Haldane was writing his appendix. This was unfortunate for two reasons.

First, Wright's paper is dominated by mathematics of remarkable originality, which cannot escape being the center of attention. Wright relied heavily on "path coefficients," a method whose relationship to other mathematical concepts and constructs was not obvious at that time. Malecot's work was needed to bring many of Wright's ideas within the reach of other mathematical theorists (Nagylaki 1989b).

Second, the manner in which Wright (1931) formulated his conclusions was curiously muffled. True, Wright's views on the population structure most favorable to evolution were put forward with perfect clarity and precision: he saw no need to change that wording during the whole of his long and productive life. These views, however, were not prominently displayed: one could easily overlook them. Moreover, in this paper, he motivated these views in terms of the need to prevent selection from depleting the variability of a species. Only a year later did his views on the rationale of population subdivision crystallize about the familiar theme of enabling a population to "shuffle" from lower to higher adaptive peaks. Finally, Wright (1931, p. 153) adduced as evidence for his views the notion of evolution as "nonadaptive branching following isolation as the usual mode of origin of subspecies, species, and perhaps even genera, adaptive branching giving rise occasionally to species which may originate new families, orders, etc."

Perhaps as a result, Haldane missed the point Wright was trying to make, focusing instead on Wright's (1931, p. 150) comment that adaptation is somewhat more effective in medium-sized populations than in very large or very small ones. A comment in Wright (1945b) suggests that Haldane was not alone in this error. Haldane also thought that Wright attached too much importance to genetic drift. Perhaps this is because Haldane foresaw that differences between related species are adaptive, as is indeed the case.

Wright's remark about differences between species usually being nonadaptive may have inflamed a dispute with Fisher, whose studies of the evolution of dominance convinced him that differences of much less than subspecific value are adaptive. This Afterword is no fit place to review a dispute that scarred population genetics, stunted its growth, and both marred the balance and blunted the perceptiveness of Fisher's vision. Those who wish to understand the dispute can consult Box (1978), Bennett (1983), and Provine (1986): no one of these three treatments suffices for a full picture.

More recently, criticisms of Wright's theory have focused on the improbability that, once an adaptive gene combination is fixed in a subpopulation, it can be exported intact to others (Haldane 1959, p. 140; Leigh 1987). These critics were sure that recombination would dissolve these new genotypes when migrants introduced them to other subpopulations. Crow, Engels, and Denniston (1989), however, show that this objection is much less compelling than had previously been thought.

The primary difficulty with Wright's theory is the difficulty of making an adequate model of it. How fast does evolution shuffle a subdivided population from peak to peak? Barton (1989) proposed a model system for analyzing this process, which was discussed in section 8: the interplay between mutation and stabilizing selection on a quantitative characteristic. This model system has many stable equilibria, with differing values of population fitness $\bar{W}$. If, at each locus, A mutates to a at a rate u, then equation (8.13) can be modified to read

$$dq_i/dt = \tfrac{1}{2}q_i(1 - q_i)\,\partial \ln \bar{W}/\partial q_i + u(1 - 2q_i)$$
$$= \tfrac{1}{2}q_i(1 - q_i)\,\partial P/\partial q_i, \tag{13.1}$$

where

$$P = \ln \bar{W} + 2u \sum_{i=1}^{m} \ln q_i(1 - q_i). \tag{13.2}$$

Now suppose that selection is acting on a finite population of effective size N, with overlapping generations, and that equation (13.1) applies to $M(q_i)$, not dq_i/dt. Then, generalizing equation (10.15), the joint probability density of the frequencies of all m A-alleles is

$$\phi(q_1, \ldots, q_m) = \frac{C \exp NP}{q_1(1 - q_1)q_2(1 - q_2) \ldots q_n(1 - q_n)}$$
$$= C\bar{W}^N[q_1(1 - q_1)q_2(1 - q_2) \ldots q_n(1 - q_n)]^{2Nu-1}$$
$$\approx C[q_1(1 - q_1)q_2(1 - q_2) \ldots$$
$$q_n(1 - q_n)]^{2Nu-1} \exp Ns[\bar{M}^2 + V]. \tag{13.3}$$

Knowing ϕ, one can calculate the rate of transition from one stable equilibrium to another, the rate of divergence between isolated populations, and the like (Barton 1989). The system is most schematic, but it is workable (barely), and in a few years' time it may allow a much more intelligent discussion of Wright's theory than I can give here.

REFERENCES

Aoki, K. 1982. A condition for group selection to prevail over counteracting individual selection. *Evolution* 36:832–842.

Artin, E. 1964. *The Gamma Function.* New York: Holt, Rinehart and Winston.

Barton, N. H. 1986a. The maintenance of polygenic variability through a balance between mutation and stabilizing selection. *Genetical Research* 47:209–216.

———. 1986b. The effects of linkage and density-dependent regulation on gene flow. *Heredity* 57:415–426.

———. 1989. The divergence of a polygenic system subject to stabilizing selection, mutation and drift. *Genetical Research* 54:59–77.

Barton, N. H., and M. Turelli. 1987. Adaptive landscapes, genetic distance and the evolution of quantitative characters. *Genetical Research* 49:157–173.

Bennett, J. H., ed. 1983. *Natural Selection, Heredity and Eugenics: Including Selected Correspondence of R. A. Fisher with Leonard Darwin and Others*. Oxford: Oxford University Press.

Box, J. F. 1978. *R. A. Fisher: The Life of a Scientist*. New York: John Wiley and Sons.

Brown, J., J. A. Cebra-Thomas, J. D. Bleil, P. M. Wassarman, and L. M. Silver. 1989. A premature acrosome reaction is programmed by mouse *t* haplotypes during sperm differentiation and could play a role in transmission ratio distortion. *Development* 106:769–773.

Bulmer, M. G. 1980. *The Mathematical Theory of Quantitative Genetics*. Oxford: Oxford University Press.

Bürger, R. 1988. Mutation-selection balance and continuum-of-alleles models. *Mathematical Biosciences* 91:67–83.

Charlesworth, B. 1988. The maintenance of transposable elements in natural populations. In *Plant Transposable Elements*, ed. O. Nelson, pp. 189–212. New York: Plenum Press.

Clayton, G. A., and A. Robertson. 1955. Mutation and quantitative variation. *American Naturalist* 89:151–158.

Coyne, J. A., and H. A. Orr. 1989. Patterns of speciation in *Drosophila*. *Evolution* 43:362–381.

Crow, J. F. 1979. Genes that violate Mendel's rules. *Scientific American* 240(2):134–146.

———. 1986. *Basic Concepts in Population, Quantitative and Evolutionary Genetics*. San Francisco: W. H. Freeman and Co.

———. 1988. The ultraselfish gene. *Genetics* 118:389–391.

Crow, J. F., and K. Aoki. 1982. Group selection for a polygenic behavioral trait: A differential proliferation model. *Proceedings of the National Academy of Sciences, USA* 79:2628–2631.

———. 1984. Group selection for a polygenic behavioral trait: Estimating the degree of population subdivision. *Proceedings of the National Academy of Sciences, USA* 81:6073–6077.

Crow, J. F., W. R. Engels, and C. Denniston. 198. Phase three of Wright's shifting balance theory. *Evolution*, in press.

Crow, J. F., and M. Kimura. 1970. *Introduction to Population Genetics Theory*. New York: Harper and Row.

Darwin, C. R. 1859. *On the Origin of Species*. London: John Murray.

Davis, P. J. 1965. The gamma function and related functions. In *Handbook of Mathematical Functions*, eds. M. Abramowitz and I. Stegun, pp. 253–293. New York: Dover Press.

Eberhard, W. G. 1980. Evolutionary consequences of intracellular organelle competition. *Quarterly Review of Biology* 55:231–249.

Endler, J. A. 1977. *Geographic Variation, Speciation and Clines*. Princeton, N.J.: Princeton University Press.

Engels,W. R. 1986. On the evolution and population genetics of hybrid-dysgenesis-causing transposable elements in *Drosophila*. *Philosophical Transactions of the Royal Society of London* B312:205–215.

Ehrman, L., and J. Probber. 1978. *Drosophila* males: the mysterious matter of choice. *American Scientist* 66:216–222.

Ewens, W. J. 1989. An interpretation and proof of the Fundamental Theorem of Natural Selection. *Theoretical Population Biology* 36:167–180.

Falconer, D. S. 1981. Introduction to Quantitative Genetics, 2d ed. London: Longmans.

Feller, W. 1968. *An Introduction to Probability Theory and Its Applications*. Vol. 1, 3rd ed. New York: John Wiley and Sons.

Felsenstein, J. 1975. Genetic drift in clines which are maintained by migration and natural selection. *Genetics* 81:191–207.

———. 1976. The theoretical population genetics of variable selection and migration. *Annual Review of Genetics* 10:253–280.

———. 1977. Multivariate normal genetic models with a finite number of loci. In *Proceedings of the International Conference on Quantitative Genetics*, eds. E. Pollak, O. Kempthorne, and T. B. Bailey, pp. 227–246. Ames: Iowa State University Press.

———. 1981. Skepticism towards Santa Rosalia, or, Why are there so few kinds of animals? *Evolution* 35:124–138.

Fisher, R. A. 1922. On the dominance ratio. *Proceedings of the Royal Society of Edinburgh* 42:321–341.

———. 1930a. *The Genetical Theory of Natural Selection*. Oxford: Oxford University Press.

———. 1930b. The distribution of gene ratios for rare mutations. *Proceedings of the Royal Society of Edinburgh* 50:205–220.

———. 1935. Dominance in poultry. *Philosophical Transactions of the Royal Society of London* B225:197–226.

———. 1938. Dominance in poultry: Feathered feet, rose comb, internal pigment and pile. *Proceedings of the Royal Society of London* B125:25–48.

———. 1950. Gene frequencies in a cline determined by selection and diffusion. *Biometrics* 6:353–361.

———. 1958a. *The Genetical Theory of Natural Selection*, 2d ed. New York: Dover Press.

———. 1958b. *Statistical Methods for Research Workers*, 13th ed. New York: Hafner.

Gillespie, J. H., and M. Turelli. 1989. Genotype-environment interactions and the maintenance of polygenic variation. *Genetics* 121:129–138.

Haldane, J. B. S. 1924. A mathematical theory of natural and artificial selection. Part II. The influence of partial self-fertilization, inbreeding, assortative mating, and selective fertilization on the composition of

Mendelian populations, and on natural selection. *Proceedings of the Cambridge Philosophical Society* 1:158–163.

———. 1932. *The Causes of Evolution*. London: Longmans Green & Co.

———. 1939. The equilibrium between mutation and random extinction. *Annals of Eugenics* 9:400–405.

———. 1948. The Theory of a Cline. *Journal of Genetics* 48:277–284.

———. 1954. *The Biochemistry of Genetics*. London: George Allen & Unwin.

———. 1959. Natural selection. In *Darwin's Biological Work*, ed. P. R. Bell, pp. 101–149. Cambridge, Eng.: Cambridge University Press.

Hamilton, W. D. 1964. The genetical theory of social behavior. *Journal of Theoretical Biology* 7:1–52.

———. 1982. Pathogens as causes of genetic diversity in their host populations. In *Population Biology of Infectious Diseases*, eds. R. M. Anderson and R. M. May. Berlin: Springer-Verlag, pp. 269–296.

Hardin, G. 1968. The tragedy of the commons. *Science* 162:1243–1248.

Harris, T. E. 1963. *The Theory of Branching Processes*. Berlin: Springer-Verlag.

Iltis, H. H. 1983. From teosinte to maize: The catastrophic sexual transmutation. *Science* 222:886–894.

Janzen, D. H. 1970. Herbivores and the number of tree species in tropical forests. *American Naturalist* 104:501–528.

Kacser, H., and J. A. Burns. 1981. The molecular basis of dominance. *Genetics* 99:639–666.

Karlin, S., and J. McGregor. 1962. On a genetics model of Moran. *Proceedings of the Cambridge Philosophical Society* 58:299–311.

Keightley, P. D., and W. G. Hill. 1988. Quantitative genetic variability maintained by mutation-stabilizing selection balance in finite populations. *Genetical Research* 52:33–43.

Kimura, M. 1962. On the probability of fixation of mutant genes in a population. *Genetics* 47:713–719.

———. 1964. Diffusion models in population genetics. *Journal of Applied Probability* 1:177–232.

———. 1965. A stochastic model concerning the maintenance of genetic variability in quantitative characters. *Proceedings of the National Academy of Sciences, USA* 54:731–736.

———. 1983a. *The Neutral Theory of Molecular Evolution*. Cambridge, Eng.: Cambridge University Press.

———. 1983b. Diffusion model of intergroup selection with special reference to evolution of an altruistic character. *Proceedings of the National Academy of Sciences, USA* 80:6317–6321.

Lande, R. 1976a. Natural selection and random genetic drift in phenotypic evolution. *Evolution* 30:314–334.

———. 1976b. The maintenance of genetic variability by mutation in a polygenic character with linked loci. *Genetical Research* 26:221–235.

Leigh, E. G., Jr. 1971. *Adaptation and Diversity*. San Francisco: Freeman, Cooper & Co.

———. 1977. How does selection reconcile individual advantage with the good of the group? *Proceedings of the National Academy of Sciences, USA* 74:4542–4546.

———. 1983. When does the good of the group override the advantage of the individual? *Proceedings of the National Academy of Sciences, USA* 80: 2985–2989.

———. 1986. Ronald Fisher and the development of evolutionary theory, I. The role of selection. *Oxford Surveys in Evolutionary Biology* 3:187–223.

———. 1987. Ronald Fisher and the development of evolutionary theory, II. Influences of new variation on evolutionary process. *Oxford Surveys in Evolutionary Biology* 4:212–263.

———. 1990. Introduccion al ambiento biotico: por que hay tantos tipos de arboles tropicales? In *Ecologia de un Bosque Tropical*, eds. E. G. Leigh, Jr., A. S. Rand and D. M. Windsor. Balboa, Panama: Smithsonian Tropical Research Institute, in press.

Leonard, J., and L. Ehrman. 1976. Recognition and sexual selection in *Drosophila*. *Science* 193:693–695.

Levins, R. 1970. Extinction. In *Some Mathematical Questions in Biology*, ed. M. Gerstenhaber, pp. 75–107. Vol. 2 of *Lectures on Mathematics in the Life Sciences*. Providence, R.I.: American Mathematical Society.

Levitt, P. R. 1978. The mathematical theory of group selection, I. Full solution of a Levins $E = E(x)$ model. *Theoretical Population Biology* 13:382–396.

Lewontin, R. C. 1962. Interdeme selection controlling a polymorphism in the house mouse. *American Naturalist* 96:65–78.

———. 1970. The units of selection. *Annual Review of Ecology and Systematics* 1:1–18.

———. 1974. *The Genetic Basis of Evolutionary Change*. New York: Columbia University Press.

Margulis, L. 1981. *Symbiosis in Cell Evolution*. San Francisco: W. H. Freeman & Co.

May, R. M., J. A. Endler, and R. E. McMurtrie. 1975. Gene frequency clines in the presence of selection opposed by gene flow. *American Naturalist* 109:659–676.

Maynard Smith, J. 1964. Group selection and kin selection. *Nature* 201:1145–1147.

Mayr, E. 1963. *Animal Species and Evolution*. Cambridge, Mass.: Harvard University Press.

Moran, P. A. P. 1962. *The Statistical Processes of Evolutionary Theory*. Oxford: Oxford University Press.

Morton, N. E., J. F. Crow, and H. J. Muller. 1956. An estimate of the mutational damage in man from data on consanguineous marriages. *Proceedings of the National Academy of Sciences* 42:855–863.

Nagylaki, T. 1975. Conditions for the existence of clines. *Genetics* 80:595–615.

———. 1989a. The maintenance of genetic variability in two-locus models of stabilizing selection. *Genetics* 122:235–248.

———. 1989b. Gustave Malecot and the transition from classical to modern population genetics. *Genetics* 122:253–268.

Ohta, T. 1988. Multigene and supergene families. *Oxford Surveys in Evolutionary Biology* 5:41–65.

Price, G. R. 1972. Fisher's "fundamental theorem" made clear. *Annals of Human Genetics* 36:129–140.

Prout, T., J. Bundgaard, and S. Bryant. 1973. Population genetics of modifiers of meiotic drive, I. The solution of a special case and some general implications. *Theoretical Population Biology* 4:446–465.

Provine, W. B. 1986. *Sewall Wright and Evolutionary Biology*. Chicago: University of Chicago Press.

Rose, M. R., and B. Charlesworth. 1981a. Genetics of life history in *Drosophila melanogaster*, I. Sib analysis of adult females. *Genetics* 97:173-186.

———. 1981b. Genetics of life history in *Drosophila melanogaster*, II. Exploratory selection experiments. *Genetics* 97:187–196.

Shami, S. A., and A. M. Tahir. 1979. Operation of natural selection on human height. *Pakistan Journal of Zoology* 11:75–83.

Simmons, M. J., and J. F. Crow. 1977. Mutations affecting fitness in *Drosophila* populations. *Annual Review of Genetics* 11:49–78.

Slatkin, M. 1973. Gene flow and selection in a cline. *Genetics* 75:733–756.

———. 1975. Gene flow and selection in a two-locus system. *Genetics* 81:787–802.

———. 1977. Gene flow and genetic drift in a species subject to frequent local extinctions. *Theoretical Population Biology* 12:253–262.

———. 1981. Pleiotropy and parapatric speciation. *Evolution* 36:263–270.

———. 1987. Heritable variation and heterozygosity under a balance between mutation and stabilizing selection. *Genetical Research* 50:53–62.

Slatkin, M., and M. J. Wade. 1978. Group selection on a quantitative character. *Proceedings of the National Academy of Sciences, USA* 75:3531–3534.

Snow, A. A., and S. J. Mazer. 1988. Gametic selection in *Raphanus raphanistrum*: A test for heritable variation in pollen competitive ability. *Evolution* 42:1065–1075.

Turelli, M. 1984. Heritable variation via mutation-selection balance: Lerch's zeta meets the abdominal bristle. *Theoretical Population Biology* 25:138–193.

———. 1985. Effects of pleiotropy on predictions concerning mutation-selection balane for polygenic traits. *Genetics* 111:165–195.

Turelli, M., and N. H. Barton. 1990. Dynamics of polygenic characters under selection. *Theoretical Population Biology*, in press.

Turner, J. R. G. 1970. Changes in mean fitness under natural selection. In *Mathematical Topics in Population Genetics*, ed. K. Kojima, pp. 32–78. Berlin: Springer-Verlag.

Watt, W. B. 1983. Adaptation at specific loci, II. Demographic and biochemical elements in the maintenance of the Colias PGI polymorphism. *Genetics* 103:691–724.

Werren, J. H., U. Nur, and C.-I. Wu. 1988. Selfish genetic elements. *Trends in Ecology and Evolution* 3:297–302.

West Eberhard, M. J. 1975. The evolution of social behavior by kin selection. *Quarterly Review of Biology* 50:1–33.

Williams, G. C. 1966. *Adaptation and Natural Selection*. Princeton, N.J.: Princeton University Press.

Wilson, D. S. 1977. Structured demes and the evolution of group-advantageous traits. *American Naturalist* 111:157–185.

———. 1980. *The Natural Selection of Populations and Communities*. Menlo Park, Calif.: Benjamin/Cummings.

Wilson, D. S., and M. Turelli. 1986. Stable underdominance and the evolutionary invasion of empty niches. *American Naturalist* 127:835–850.

Wright, S. 1931. Evolution in Mendelian populations. *Genetics* 16:97–159.

———. 1932. The roles of mutation, inbreeding, crossbreeding and selection in evolution. *Proceedings of the Sixth International Congress in Genetics*, vol. 1, pp. 356–366.

———. 1934. Physiological and evolutionary theories of dominance. *American Naturalist* 68:25–53.

———. 1935. Evolution in populations in approximate equilibrium. *Journal of Genetics* 30:257–266.

———. 1937. The distribution of gene frequencies in populations. *Proceedings of the National Academy of Sciences, USA* 23:307–320.

———. 1945a. The differential equation of the distribution of gene frequencies. *Proceedings of the National Academy of Sciences, USA* 31:383–. 389.

———. 1945b. Tempo and mode in evolution: A critical review. *Ecology* 26:415–419.

NOTES TO ORIGINAL TEXT BY E. G. LEIGH, JR.

1. In his original text Haldane has exp $n/4$, not exp $n/8$. However, if one substitutes $x = \frac{1}{2}\sqrt{n}$ into the equation for p and applies the approximation cited in equation (3.10) of the Afterword, one obtains the following equation:

$$p = 1/(\sqrt{n\pi/2}\ e^{n/8}).$$

This is academic: Fisher's approximation for p only applies when r/d is far smaller than 1. (See E. G. Leigh, Jr. *Oxford Surveys in Evolutionary Biology* Vol. 4, p. 257.)

2. In the last equation, Haldane has σ in the denominator (see Afterword); it should be σ^2.

3. Haldane expressed y_n as $1/(1 + e^{\frac{1}{2}kn})^2$, as if a were recessive, and gives the equilibrium gene ratio as $(2K/k - 1)$, which is the *unstable* equilibrium between selection $\frac{1}{2}k$ for A is gametes and

selection K for aa among diploids, a circumstance where A's frequency q obeys the equation

$$dq/dt = (k/2)q(1 - q) - Kq(1 - q)^2.$$

A corrected version of this paragraph is:

> Where selection operates on the gametes of one gender, e.g. pollen tubes, favouring A gametes we have $kn = 2 \log_e u_n$, or $u_n = e^{\frac{1}{2}kn}$ (letting A be recessive). The proportion of recessives is $y_n = e^{kn}/(1 + e^{\frac{1}{2}kn})^2$, so when recessives are few they increase or decrease in geometrical progression, as do dominants when these are few. Hence selection of this type will begin operating at once on a new and therefore rare recessive gene. The fact that the expression of this gene in the diploid phrase is disadvantaged will not begin to stop its spread until it is fairly common. If K be the coefficient of selection (against AA) in the diploid, equilibrium is reached when $u = k/(2K - k)$. If competition is only between gametes from the same individual, as is commonly the case, the rate of selection is halved, but the phenomenon is qualitatively similar.

4. Haldane omits the factor 1/3 in the two following equations:

$$u_{n+1} - u_n = u_n \left(\frac{2k_1}{u_n + 1} + k_2 \right) \Big/ 3;$$

$$(2k_1 + k_2)n/3 = \log_e \left(\frac{u_n}{u_o} \right) + \frac{2k_1}{k_2} \log_e \frac{u_n + 1 + \dfrac{2k_1}{k_2}}{u_o + 1 + \dfrac{2k_1}{k_2}}$$

To see why it should be there consider an allele A on the X chromosome. If there is no selection, A's frequency $P(n + 1)$ among males of generation $n + 1$ is its frequency $Q(n)$ among the females of generation n, as males get their X chromosomes from their mothers. A's frequency $Q(n + 1)$ among the females of generation $n + 1$ is $\frac{1}{2}[P(n) + Q(n)]$, as females set one X each from father and mother. Thus

$$2Q(n + 1) + P(n + 1) = 2Q(n) + P(n) = 2Q(0) + P(0).$$

The sum $2Q + P$ is invariant. As sexual reproduction ensures that Q and P approach each other, A's frequency approaches $[2Q(0) + P(0)]/3$ in both sexes.

5. This equation should read:

$$kn = \log_e\left(\frac{u_n}{u_o}\right) + \frac{(2 - 2\lambda)}{\lambda} \log_e\left(\frac{2 - \lambda + \lambda u_n}{2 - \lambda + \lambda u_o}\right)$$

6. This should read:

$$1 + (\lambda k)/(2 - \lambda).$$

7. Here, Haldane's k is the difference in fitness between recessive and dominant phenotypes, so $du/dn = ku/(1 + u)$.

Index